DEDICATION

This book is dedicated to the deeds and exploits, to the sacrifices and achievements of the generation of Australians who proudly served in the RAAF's Boston, Mitchell and Liberator Squadrons.

BOSTON, MITCHELL AND LIBERATOR

IN AUSTRALIAN SERVICE

by Stewart Wilson

ACKNOWLEDGEMENTS

As has been the case with the previous seven books in this series, a great deal of assistance has been given to the author by people interested in recording aspects of the history of the RAAF.

Firstly, Bob Livingstone, who allowed the author free access to material he'd researched which covered American Liberators in the Java campaign, No 22 Squadron's historian Dick Hutchinson, David Wilson of the RAAF's Historical Section, Bob Piper (formerly of RAAF Historical), Neil Mackenzie, Keith Hall, Mike Kerr, Flt Lt John Bennett, Nelson Hill and the artists who provided the cover paintings and drawings, Harry Margulies and John Allen.

Thanks also to production manager Maria Davey (who worked through ill health to get the book finished) and the publishers, Jim and Margaret Thorn, who are not only my 'employers' but also good friends.

Finally, my thanks to the buyers of these books. If they were not commercially viable they would not be published, and the ongoing support of the readers is most gratifying, particularly as we now offer the eighth title in this series.

STEWART WILSON
Sydney, 1992

Published by Aerospace Publications Pty Ltd (ACN: 001 570 458), PO Box 3105, Weston Creek, ACT 2611, publishers of monthly Australian Aviation magazine.
Production Manager: Maria Davey

ISBN 1 875671 00 5

Proudly printed in Australia by Pirie Printers Sales Pty Ltd, 140 Gladstone Street, Fyshwick 2609
Distributed by Network Distribution Co, 54 Park Street, Sydney 2000
Film by Pirie Printers Sales Pty Ltd, Canberra
Distribution in North America by Motorbooks International, 729 Prospect Ave, Osceola, Wisconsin 54020, USA. Fax: 715 294 4448

CONTENTS

FRONT COVER: Douglas Boston III A28-13/B *Rocky's Girl* of No 22 Squadron, Ward's Strip, Port Moresby, late 1942; North American B-24J Mitchell A47-26/KO-K of No 2 Squadron, late 1944; Consolidated B-24L Liberator A72-84/GR-D *Cock o' the North* of No 24 Squadron, late 1944. (Artwork Zvi Margulies)
(Drawings not to scale)

Aerospace Publications Pty Ltd wishes to advise purchasers of this book that the lack of colour illustrations is due to the fact that no suitable material from the period covered was available. Our normal policy of running a large variety of colour photographs will continue with future books in the Airpower Collection, where they are available.

INTRODUCTION

The eighth book in our 'Australian Air Power' series deals with three strike/bomber types which served the RAAF during World War II – the Douglas Boston, North American Mitchell and Consolidated Liberator.

All three types served the RAAF with distinction, although in the case of the Boston and Mitchell, in quite small numbers.

Sixty-nine Bostons were delivered to the RAAF between 1942 and 1944 and the type served with just one squadron, No 22. Despite this, the aircraft achieved a fame which belies its numbers. It was in a Boston that the RAAF's only Victoria Cross of the Pacific war was awarded to Flt Lt W E Newton. For this reason alone, the Boston holds a special place in the history of the RAAF and Newton's VC reflects the substantial contribution No 22 Squadron and its Bostons made to the Allied campaign in the New Guinea area.

The Mitchell was also delivered to the RAAF in modest quantities – just 50 – and served with only one squadron. No 2 Squadron received its first Mitchells in 1944 and operated them against enemy targets to the north of Australia, proving particularly effective against shipping.

But the RAAF's and Australia's connection with what is generally regarded as being the best medium bomber of the war actually began two years earlier in 1942 with the equipping of No 18 Netherlands East Indies Squadrons with the type. No 18 Squadron operated under RAAF command (with a mix of Dutch and Australian air and ground crews), initially from Canberra and then the Northern Territory where it was later joined by No 2 Squadron's Mitchells from 1944 as part of No 79 Wing RAAF. Together, the two Mitchell squadrons were an effective team.

By contrast, the Liberator was operated in large quantities by the RAAF, some 287 being delivered in 1944-45. Seven squadrons operated the type and those of No 82 Wing (Nos 24, 21 and 23) in particular playing a vital role in the Borneo campaign. Before the delivery of the RAAF's own Liberators had begun, training of Australian aircrews was carried out in conjunction with USAAF Liberator units, Australians flying into combat in the American aircraft.

Unlike the Boston and Mitchell, the Liberator continued in RAAF service for a time after the war had ended, albeit briefly, and by 1948 the big bomber had been retired from service.

ROYAL AUSTRALIAN AIR FORCE

PER ARDUA AD ASTRA

Four of No 22 Squadron RAAF's Douglas Boston IIIs captured by the photographer during the return flight to base after a raid on Lae. The date is 22 February 1943 and the first four aircraft are A28-7/DU-H, A28-15/DU-Q, A28-8/DU-J and A29-6/DU-G. (via Mike Kerr)

DOUGLAS
A-20 BOSTON

DOUGLAS BOSTON

The Douglas A-20 Havoc (or Boston as the British called it) series of aircraft was in many ways one of the more successful but also largely unsung workhorses of the Second World War. It never achieved the public fame of one or two of its light bomber/strike contemporaries yet served in all the major theatres of the war in the colours of ten nations. Production reached the healthy total of 7,385 by the time it ended in September 1944, 380 of these under licence by Boeing and the remainder by the parent company. Of the aircraft produced, more than 3,100 were supplied to the Soviet Union.

The Boston was a versatile aircraft which incorporated many features which could be considered advanced for its time, particularly when compared to European aircraft of the immediate pre-war period. It featured the then novelty of tricycle undercarriage, and was more generally complex than the norm. Even in its earlier versions its performance was excellent with a top speed comfortably in excess of 300mph (500km/h) combined with a bomb load which began at 2,000lb (907kg) in the early versions and doubled in later models. An innovation was the incorporation of a bomb bay split into two compartments.

As it developed, the aircraft's gun armament increased and was used more as offensive weapons rather than for defensive purposes, while other variants were used as night fighters and night intruders. The Royal Air Force dubbed its night fighter Bostons 'Havocs', the name by which the aircraft became generally known in the USA. In Australian service, the name Boston was used.

In view of the assertion that the Boston's contribution to the Allied war effort was undoubtedly understated, it's interesting to note that in Australian service, its reputation and the reporting of its deeds was probably disproportionate to the other extent. Only one Royal Australian Air Force Squadron – No 22 – operated the Boston and just 69 were taken on charge between March 1942 and October 1944. Yet, due to some extraordinary exploits by 22 Squadron in the South West Pacific Area, the Boston was well known by the public.

Only three RAAF airmen were awarded the Victoria Cross in World War II. One of them was Flt Lt W E Newton, a 22 Squadron pilot who was posthumously awarded his VC in March 1943. This alone puts the Boston into a special category of Australian service.

The Douglas Model 7

The United States of America of the 1930s was not a wonderful place to be for those in the business of designing and manufacturing military aircraft. The country's policies were isolationist and it was apparently unaware of (or chose to ignore) the war clouds which were gathering in Europe and the Far East. The US was not alone in that attitude in the mid 1930s, although other nations were quicker to realise the threat and began to rearm.

Although the USA was officially disinterested in potential problems on the other side of the world, some individuals were concerned by them, and the general lack of modern equipment under development for the US services. Like so many of the more successful military aircraft throughout the course of aviation's short history, what ultimately emerged as the Douglas A-20/Havoc/Boston series began as a private venture.

The Douglas Aircraft Company realised as early as the end of 1935 that the United States Army Air Corps' typical single engined attack aircraft (such as the Northrop A-17) would not be adequate in a modern war. The company therefore set about designing a twin engined light attack bomber without the support of a request from the army. Recent international events had reinforced Douglas' perception that things were moving towards a period of uncertainty: Japan had invaded Manchuria, Mussolini's Italy had invaded Ethiopia, Adolf Hitler had established Germany's *Luftwaffe* in direct defiance of the Treaty of Versailles and had reoccupied the Rhineland and – in mid 1936 – the Spanish Civil War would begin.

Detail design work on the Douglas Model 7A began in March 1936 under the direction of two of the USA's most famed designers, Jack Northrop and Ed Heinemann. Northrop had previously become well known for several designs and was well versed in all metal stressed skin construction techniques, a feature which had been applied to several of his designs in previous years. His Northrop Aviation Corporation had recently merged with Douglas when Model 7 work began.

Heinemann would go on to be associated with many well known Douglas designs. The company itself had been in existence since 1921 and like Northrop, had based its operations around California. Douglas was a large company by the mid and late 1930s with numerous successful designs to its credit. By the time design work on the Model 7 was underway, the company had already developed expertise with modern design methods and all metal structures, notably through the DC-1, DC-2 and recently flown (December 1935) DC-3 series of airliners, the latter revolutionising air transport in both civil and military guises and going on to become one of the genuinely great aircraft of all time.

On the military side, Douglas' most recent designs of modern specification were the single engined TBD Devastator naval torpedo bomber (first flight April 1935) and the operationally unsuccessful B-18 Bolo twin engined bomber, which combined the DC-2/DC-3 wings, tail surfaces and engines with a new fuselage. Both of these aircraft were of all metal stressed skin construction and featured retractable undercarriages.

Without a formal USAAC requirement on which to base their work, Northrop and Heinemann were groping in the dark to some extent but decided to come up with a multi engined aircraft which fitted into the general category of 'attack'. The earliest considerations assumed the aircraft could be used for several different roles, whether they be conventional medium altitude bombing, observation/ reconnaissance or low level strike duties including strafing.

Early proposals were called the Douglas Model 7A and emerged as a shoulder wing aircraft powered by two 450hp Pratt & Whitney R-985 Wasp Junior nine cylinder radial engines. A maximum speed of 250mph (400km/h) was estimated and the aircraft was designed to carry a crew of three. Maximum weight was 9,500lb (4,309kg) and the bomb load was set at 1,000lb (454kg). Gun armament was six 0.30in machine guns and an interesting concession to the Model 7A's 'multi-role' nature was an interchangeable bomb bay and observation compartment.

Design work on the Model 7A continued through 1936 and was about 50 per cent complete by the end of the year when it ground to a halt pending the issue of a specification or design competition by the USAAC following the latter's evaluation of intelligence gathered from the Spanish Civil War and other skirmishes of the recent past.

It wasn't until the northern hemisphere autumn of 1937 that the Army did issue its requirements, and a design competition was held. The basic requirements of the specification were a maximum speed of over 200mph (320km/h), a 1,200lb (544kg) bomb load and a range of 1,200 miles (1,930km).

The USAAC competition prompted a redesign of the Model 7A under the designation Model 7B. Jack Northrop had meanwhile resigned from Douglas at the end of 1937 to re-establish his own aircraft manufacturing company, leaving Heinemann in charge of the programme. The result was a much more powerful and heavier aircraft. High winged and sitting on a tricycle undercarriage, the 7B was powered by a pair of 1,100hp Pratt & Whitney R-1830 Twin Wasp 14 cylinder radials, maximum weight was calculated at 15,000lb (6,804kg) and armament comprised the prescribed bomb load plus retractable dorsal and ventral turrets each housing a single 0.30in machine gun.

In addition, a pair of 0.30in machine guns were planned for installation on each side of the nose for ground strafing and an innovative feature was the option to replace the standard plexiglass navigator/bomb aimer's nose section with a metal covered 'solid' one containing two extra 0.50in machine guns for dedicated ground attack missions. The plan was to have these alternative noses detachable as a unit and swapped around as required.

Douglas was not the only manufacturer to submit a design for the USAAC competition. Bell, Stearman and Martin were also contesting it and each company was invited to build a prototype in July 1938. Bell withdrew at this stage and was replaced by North American, which built and flew its NA-40, which then evolved into the famous B-25 Mitchell medium bomber. The Martin proposal was developed as the Maryland.

Construction of the Model 7B began immediately, and by late October it was ready for flight. This was achieved on the 26th of the month. The 7B immediately impressed with its speed and manoeuvrability. It flew with both the 'solid' and 'bombardier' noses during its short life and the only major modification applied was the moving of the horizontal tail surfaces forward and giving them some dihedral.

With hindsight it's interesting to note the aircraft looked very much like a larger version of the Aero Commander series of business twins which first appeared in the late 1940s. This is no coincidence, as the Aero Commanders were conceived in the final year of World War II by two of Douglas' employees, who left to set up their own company after Douglas showed little interest in their design for a light twin. One of them was Ted Smith, who had worked as a senior member of the A-20 design team during the war years. As well as the Aero Commander family, Smith was also responsible for the design of the Aerostar series of twin engined light aircraft in the 1960s.

The French Connection

The first flight and early test flying of the Model 7B coincided with a visit to the USA by representatives of the French Purchasing Commission, who were looking for modern combat aircraft with which to re-equip the *Armée de l'Air* in the wake of the Munich Crisis and the obvious signals of Hitler's intentions. The new Douglas design was of interest to them, although the USA's isolationist policies of the time meant that although official approval was given to the French to inspect and watch the 7B go through its paces, this was all kept under wraps. For Douglas, the chance to sell the aircraft to France was not to be missed, because no orders had as yet been forthcoming from the USAAC.

The secrecy surrounding the French involvement and the Douglas Model 7B both came to an end on 23 January 1939. During a test flight, the aircraft crashed on landing, killing the pilot John Cable. Cable's passenger – ostensibly an observer/engineer – survived, but it was soon revealed that he was in fact Captain Maurice Chemidlin, a member of the French Purchasing Commission. The uproar was heard all over the United States, the newspapers having a field day in concert with the very strong isolationist lobby. Even President Roosevelt had to take personal blame and apologise to the nation.

Despite all this, the French were very happy with what they had seen in the 7B and immediately ordered 100 examples of an improved version, the DB-7. ('DB' for 'Douglas Bomber', as 'DC' stood for 'Douglas Commercial').

The French required more bomb load, armour and range (along with French instruments and machine guns) in their aircraft, giving Ed Heinemann the opportunity to incorporate new design features which resulted in the definitive A-20/Havoc/Boston shape. Externally, the unchanged wing was moved down to a shoulder position, the fuselage itself was made narrower but considerably deeper. The 1,100hp Twin Wasp engines were retained but now mounted under rather than on the centre of the wing, the undercarriage was modified and the structure generally strengthened.

Maximum weight increased to 15,150lb (6,872kg), the maximum bomb load to 2,080lb (943kg) with which the DB-7 could fly 462 miles (743km) or twice that far with reduced load. Maximum speed was 280mph (450km/h) at sea level or 305mph (491km/h) at 9,650 feet, the critical height of two R-1830-SC3G engines with single speed superchargers. The DB-7 was built only with a glass nose section but there were four fixed Chatellerault 7.5mm machine guns on the lower nose and single flexibly mounted similar guns in ventral and dorsal rear fuselage positions.

Douglas worked with tremendous speed when the French order was received and when the new design was completed. The DB-7 recorded its first flight on 17 August 1939 and almost incredibly, the first production example was handed over to the French from Douglas' El Segundo (California) plant

Two views of the predecessor of the A-20/Boston/Havoc series, the Douglas Model 7B. First flown in October 1938, the single Model 7B was powered by two 1,100hp Pratt & Whitney R-1830 Twin Wasps. Although similar in general configuration to the aircraft which followed, many substantial differences are evident.

The Douglas DB-7 first flew in August 1939 and was developed to a French requirement. This aircraft featured the definitive Boston airframe basic design but was still powered by R-1830 engines. Following the fall of France, large numbers were taken over by the Royal Air Force as the Boston I and II. Many were converted to 'Havoc' night intruder (as illustrated here) and night fighter configuration. Note the tapered fin and rudder design. (via Neil Mackenzie)

just two months later, in October. In that same month the French ordered another 170 to be equipped with 1,200hp R-1830-S3C4-G Twin Wasps with two speed superchargers and improved performance at altitude, including a top speed of 322mph (518km/h) at 15,200 feet. Suddenly, Douglas found itself with a healthy orderbook for its new attack bomber, because the previous June a further 63 examples had been ordered by the USAAC as the A-20.

The Upgraded DB-7A

The USAAC order began the process that would see the aircraft's definitive powerplant installed, the Wright R-2600 Cyclone 14 cylinder two row radial. Produced in numerous variants, the R-2600 was rated at 1,200 horsepower in its early versions, increasing to 1,600/1,700hp as it was applied to later Bostons and A-20s.

In October 1939 the French also expressed interest in an R-2600 engined version, the extra power providing improved performance. Under the factory designation DB-7A, the new variant was powered by two 1,600hp Wright R-2600-A5B engines fitted with two speed superchargers. Structurally, the DB-7A differed little from its lower powered predecessor, although it did feature a taller fin and rudder of broader chord to increase directional stability. One of the French DB-7s had already been experimentally fitted with twin endplate fins and rudders but these were discarded for the enlarged single unit.

Other major changes were mainly in the area of the engine cowlings and nacelles, which were longer. The intake and exhaust designs were different and the cooling gills were modified. In addition, a modified undercarriage design (originally developed for the A-20) was incorporated and there was some local structural strengthening to cope with higher weights and the extra power. The French ordered 100 DB-7As in October 1939 and by the time they capitulated to the Germans in June 1940, some 951 were on order for the *Armée de l'Air*. By this time, some 65 DB-7s were in French hands and another 130 or so were in transit. As is discussed below, some of these and other aircraft intended for France were diverted to the Royal Air Force.

The DB-7A featured a further increase in maximum weight to 16,700lb (7,575kg) and a maximum speed of 308mph (495km/h) at sea level and 344mph (553km/h) at 12,500 feet, substantial performance for the day and only slightly

The DB-7A, a minimal redesign of the DB-7 to accommodate more powerful (1,700hp) Wright R-2600 engines. The larger fin and rudder and modified engine nacelles should be compared with the DB-7. Once again, the DB-7A was developed to the requirements of the French but most ended up in RAF service, converted to Havoc II night fighters with a solid nose containing 12 0.303in machine guns.

W8268 was one of the Royal Air Force Boston IIIs converted to intruder specifications for service with No 418 Squadron. (via Neil Mackenzie)

Although externally similar to the DB-7A, the DB-7B (RAF Boston III) in fact featured a substantially redesigned and strengthened structure allowing increased weights.

slower than the contemporary Spitfire Mk.I and Messerschmitt Bf.109E single engined fighters.

At this point it might be worth trying to sort out some of the potentially confusing designations applied to the family of aircraft. DB-7 was the factory designation of the aircraft; A-20 was the US Army Air Corps designation ('A' for 'Attack'); the British called the aircraft the Boston and the Havoc when converted for night fighting duties, a name which was also applied to later USAAC (or US Army Air Force from June 1941) models. In Australian service, the aircraft was normally referred to as the Boston.

Bostons For The RAF

Despite Douglas' enormous effort in building the entire first batch of French DB-7s by the end of 1939, it took considerable time to get the aircraft into active service. After handing over to the French in the USA, the aircraft were shipped to Casablanca where they were reassembled and test flown before being delivered to conversion units in French North Africa and France. As mentioned above, only 65 were in France by the time of the capitulation and as a result the aircraft played only a minor role in the Battle for France. The first DB-7 sortie against the Germans was on 31 May 1940 when 12 aircraft attacked enemy columns near St Quentin. Just over three weeks later, France surrendered to the Germans.

The result was that the RAF took over large numbers of undelivered French aircraft, plus a small number intended for Belgium. Aircraft from the original batch of French DB-7s with single speed supercharged R-1830 Twin Wasps were called the Boston I, and those with two speed superchargers were dubbed Boston IIs. The latter were refitted with British instrumentation and guns. The first diverted Bostons arrived in Britain during August 1940.

At that stage the Battle of Britain was nearing its climax and the RAF's lack of night fighters and intruders resulted in more than 100 Boston IIs being converted to these roles. Initially known as the 'Ranger' and finally the Havoc I, aircraft converted for the night intruder role were fitted with flame damping exhausts and proved extremely useful in their intended role, causing considerable damage to many targets, especially German airfields. Operating at night and at low altitude, they were difficult to intercept and carried a hefty punch with their 2,000lb (907kg) bomb load and battery of four 0.303in machine guns in the nose.

More heavily modified were the Havoc I night fighters. These carried AI Mk.IV radar and an extra battery of four machine guns in their solid nose. Again, this was a successful conversion and the theme was carried a step further when the Wright R-2600 powered DB-7As originally intended for France began arriving. These aircraft were fitted with a further revised solid nose housing no fewer than 12 0.303 machine guns and were designated the Havoc II.

Most of these aircraft were further converted to Havoc II Turbinlite configuration, an interesting concept in night fighting consisting of the installation of a 2,700 million candlepower searchlight in the nose, fed by large and heavy batteries installed in the bomb bay. The Turbinlite Havocs could carry no armament, the theory of their operation revolving around operating in concert with a pair of Hawker Hurricanes. The Havoc would pick up a target with its radar and when within range, illuminate it for the Hurricanes.

In practice, the idea was not successful and the ten operational squadrons which had been formed around the two types of aircraft were disbanded in early 1943 when more effective Beaufighter and Mosquito night fighters became readily available.

The British had meanwhile ordered their own version of the DB-7A back in February 1940, placing an initial order for 150 and increasing it to 300 three months later. Under the factory designation DB-7B and RAF name Boston III, this variant was externally similar to the French DB-7A but featured British instruments and guns (four fixed 0.303in Browning machine guns in the nose, two of them in streamlined blisters, twin flexibly mounted 0.303s in the dorsal position plus a single ventral gun), a substantially redesigned and strengthened structure which allowed an increase in gross weight to 20,320lb (9,217kg) and an increase in fuel capacity to 329 imperial gallons (1,495 l) accommodated in self sealing wing tanks.

Boeing manufactured 380 Bostons under licence in 1941-42 comprising 240 Boston IIIAs for the RAF and 140 similar A-20Cs for the USAAF. Eleven of the Boston IIIAs found their way into RAAF service. (via Neil Mackenzie)

The initial Douglas DB-7 variant for the USAAC was the A-20, the first of 63 being handed over in mid 1940. Structurally similar to the British DB-7B/Boston III, the A-20 differed in featuring American guns and equipment plus turbosupercharged Wright R-2600 engines, with the turbochargers mounted on the outer flanks of the nacelles, as illustrated in the lower photograph. Considered unnecessary (and difficult to cool), the turbocharged engine was dropped on all subsequent variants.

The split bomb bay was lengthened slightly to accommodate special British ordnance and the plexiglass nose was six inches longer and featured modified side contours in which the previous stepped arrangement was replaced with a straight diagonal design. Powerplants remained 1,600hp Wright R-2600-A5Bs with two speed superchargers and there was provision for four crew members – pilot, bomb aimer and two gunners.

The first DB-7B Boston III first flew in January 1941 and deliveries to the RAF began the following May. Three were subsequently converted to Turbinlite configuration and others to intruder specification similar to the earlier Havoc Is and with a belly pack of four 20mm cannon.

The British designation Boston III was also applied to a handful of ex-French DB-7As which had not been converted to Havoc II configuration, while the same designation was given to aircraft from yet another frustrated French order. Originally called the DB-7C and then DB-73, these aircraft had been ordered by France in May 1940 but with its fall they were completed to British DB-7B standards and delivered as Boston IIIs. The Boston IIIA was simply the standard aeroplane manufactured by Boeing and with minimal modifications (electrical system, removal of flame damping exhausts, extended 'tropical' carburettor intake on top of cowling), delivered between October 1941 and March 1942.

Hot and dusty in the North Africa/Mediterranean theatre of operations, this Boston III displays some cowling/rear fuselage detail as well as the sideways hinging cockpit cover. (via Neil McKenzie)

By this time A-20/Boston production was in full swing with aircraft coming off Douglas' El Segundo, Long Beach and Santa Monica production lines, plus Boeing's Seattle plant.

By early 1942 the RAF had received some 1,050 Bostons including 160 DB-7/ Boston Is, 99 DB-7A/Boston IIs, 541 DB-7B/Boston IIIs and 240 Boeing built Boston IIIAs. Many of the original 100 *Armée de l'Air* DB-7s fell into Vichy French hands in North Africa in 1940 but they were used sparingly. Their only major action was a strike against Gibraltar.

America's A-20s

Throughout the somewhat confusing period of the French and British DB-7 acquisitions and diversions, development of the aircraft for the USAAC/ USAAF was continuing. The first was the A-20, built to fill the original USAAC order for 63 aircraft placed in June 1939. The first aircraft was delivered in mid 1940 and was structurally similar to the DB-7B Boston III. It differed in having American calibre (0.30in) machine guns and general items of equipment, but its major difference was in the powerplant. The 1,700hp Wright R-2600-7 engines were fitted with turbochargers which were mounted internally on the outer flanks of the engine nacelles, changing the shape of these considerably.

The turbocharger installation proved to be difficult to cool and as the A-20 was considered to be a low/medium altitude aircraft, it was also considered unnecessary. Sixty of the 63 A-20s were therefore converted to P-70 night fighters with unsupercharged R-2600-11 engines, British AI radar installed in a solid nose and a pack of four 20mm cannon mounted under the bomb bay. The P-70 had poor altitude performance and most were used for training crews in radar controlled interception procedures. The three remaining A-20s were converted to F-3 armed photo reconnaissance aircraft.

The first attack version of the DB-7 family to be delivered to the USAAC was the A-20A. Except for the calibre of its guns and the installation of American equipment, this variant was similar to the RAF's Boston III. Turbocharged engines were now a thing of the past in the DB-7 family and the A-20A was fitted with two 1,600hp R-2600-3s with two speed superchargers. Provision was made for the installation of a single remotely controlled machine gun in the rear of each engine nacelle, something which had been experimented with in the F-3 conversions. A-20A production reached 143 examples in 1940.

Despite being produced in 1941, the A-20B was in some ways inferior to the A-20A which had preceeded it, being based on the DB-7A airframe rather than the DB-7B and as a result featuring no fuel tank self sealing and a comparative lack of armour. This helps explain why 665 of the 999 examples produced went to the Soviet Union under Lend-Lease. Other differences included the use of 1,690hp R-2600-11 engines, the installation of a pair of 0.50in machine guns in the nose in place of the four 0.30s and another 0.50in gun in the dorsal position. The nose glazing was of the stepped rather than diagonal style and provision was made for the installation of auxiliary fuel tanks. The A-20s nacelle guns were retained despite their not being very successful.

The XA-20B was an A-20A modified to have three powered turrets installed

THE DOUGLAS A-20

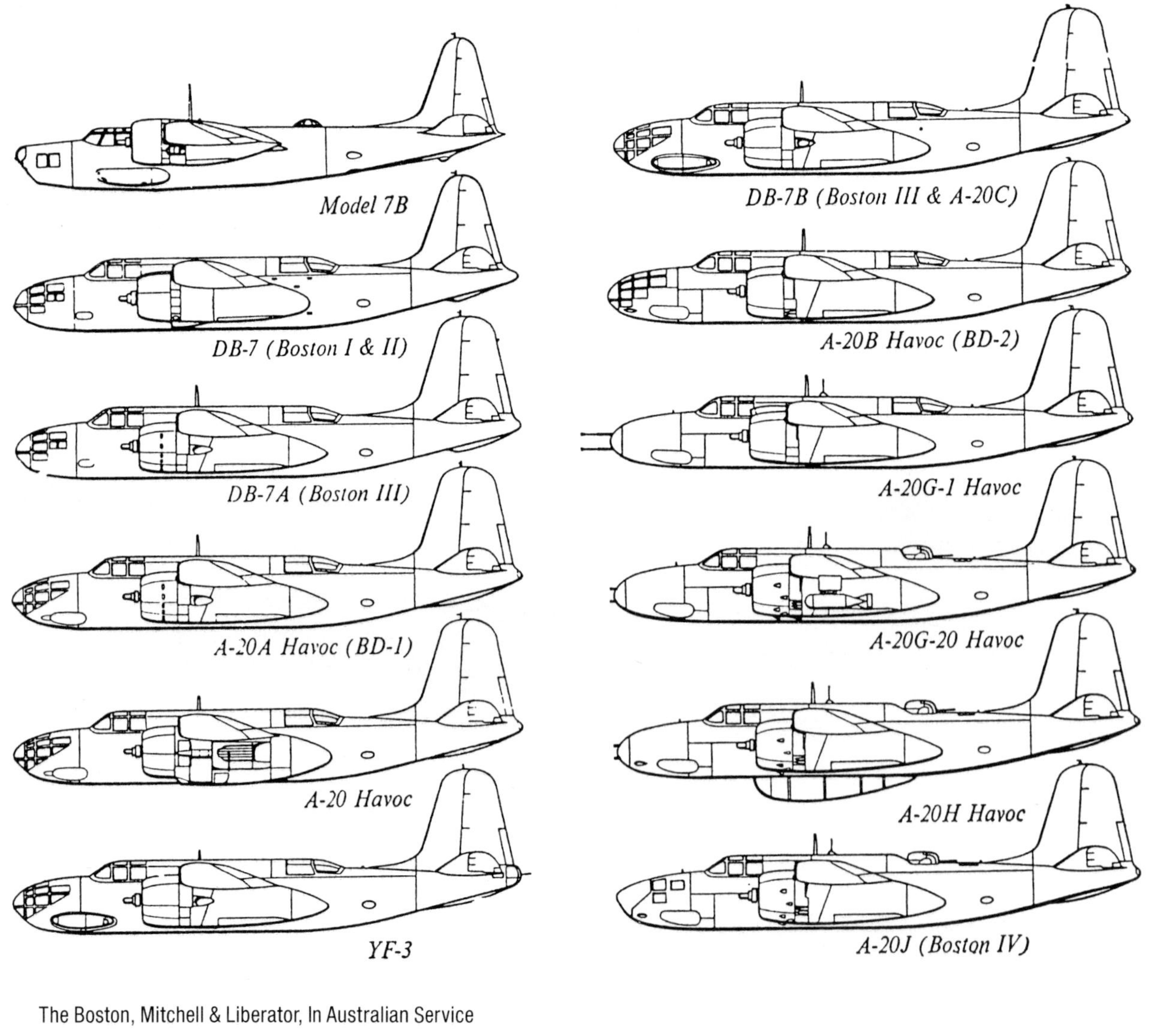

Of the 63 A-20s built, 60 were converted to P-70 night fighters with the turbochargers removed, British radar installed and a pack of four 20mm cannon mounted in the nose.

The first Havoc attack variant delivered to the USAAC was the A-20A, similar to the RAF's Boston III except for the fitting of American guns and equipment. One hundred and forty-three were delivered in 1940.

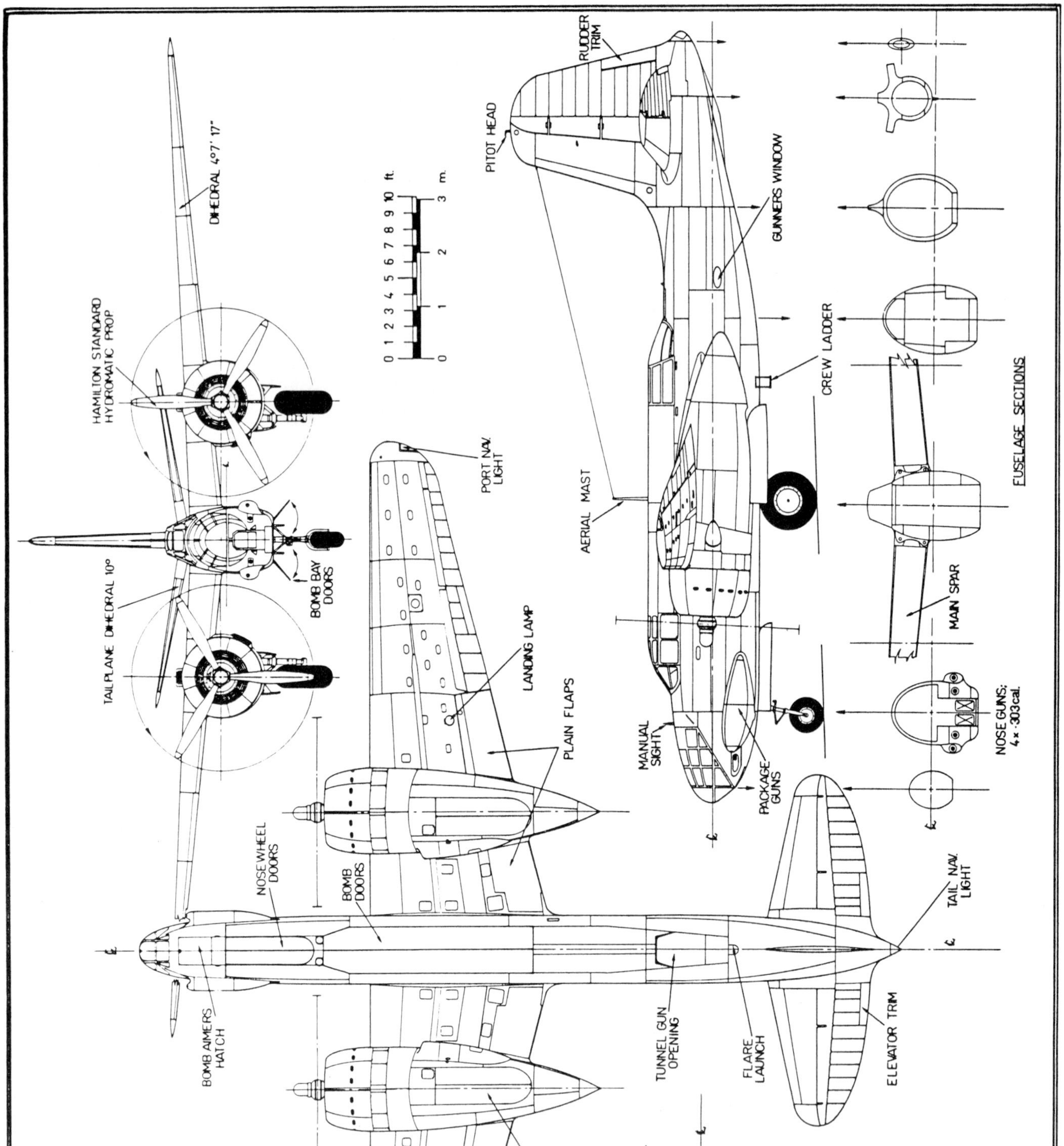
DOUGLAS BOSTON MK III
© J.F.ALLEN 1992
HAMILTON STANDARD HYDROMATIC PROP
DIHEDRAL 4°7' 17"
TAILPLANE DIHEDRAL 10°
BOMB BAY DOORS
0 1 2 3 4 5 6 7 8 9 10 ft.
0 1 2 3 m.
PORT NAV. LIGHT
PITOT HEAD
RUDDER TRIM
GUNNERS WINDOW
CREW LADDER
AERIAL MAST
LANDING LAMP
PLAIN FLAPS
MANUAL SIGHT
PACKAGE GUNS
FUSELAGE SECTIONS
MAIN SPAR
NOSE GUNS: 4 x ·303cal.
NOSEWHEEL DOORS
BOMB DOORS
BOMB AIMERS HATCH
TUNNEL GUN OPENING
FLARE LAUNCH
ELEVATOR TRIM
TAIL NAV. LIGHT
MAIN GEAR DOORS
MAINTENANCE PANELS
AILERON TRIM
EXTERNAL ARMOUR
* SCRAP VIEW: NOSE SECTION A-20G
* SCRAP VIEW: NOSE SECTION, A-20G
4 x ·50 cal.
2 x ·50 cal.

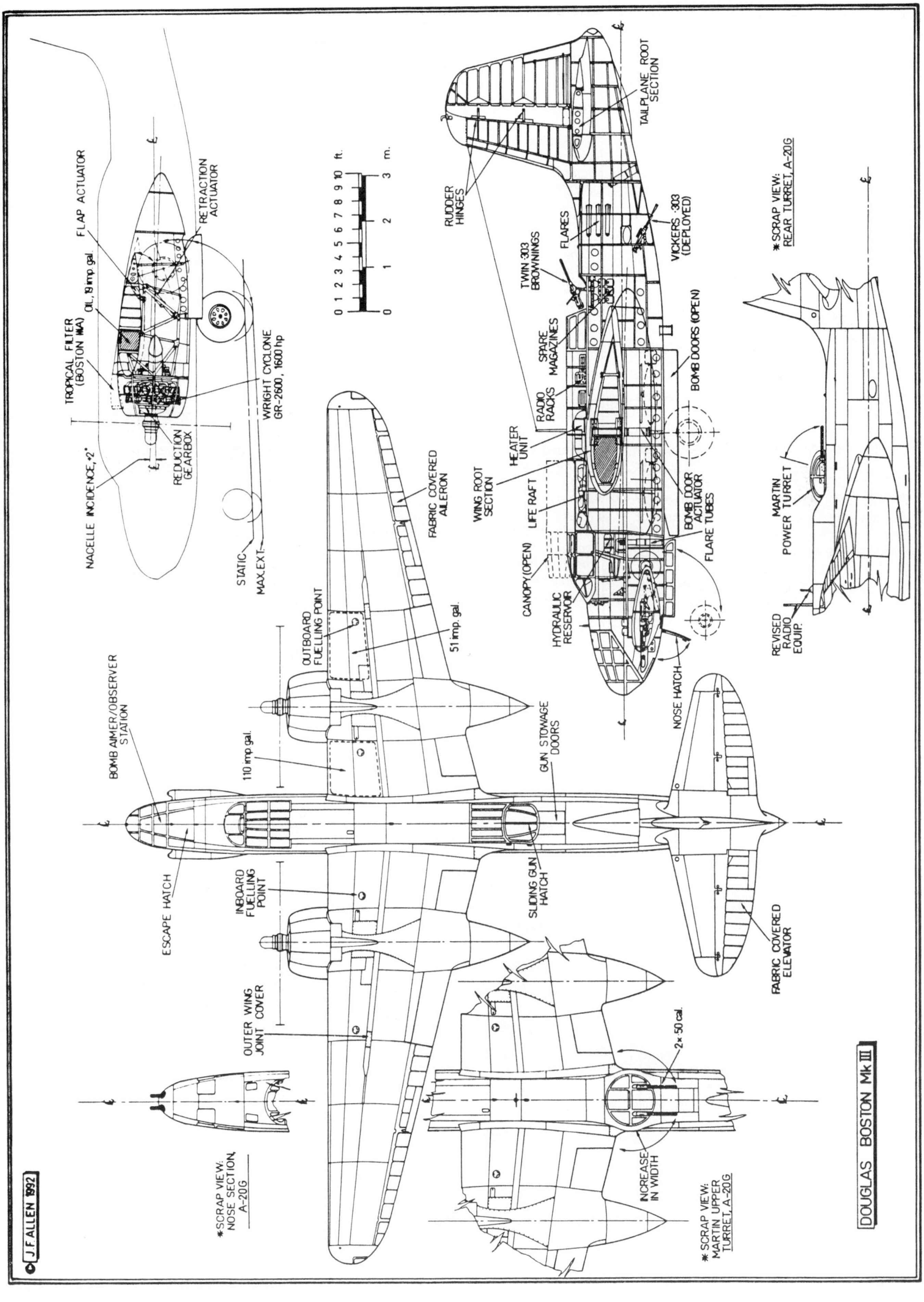

J.F. ALLEN 1992
DOUGLAS BOSTON Mk III
NACELLE INCIDENCE, +2°
REDUCTION GEARBOX
TROPICAL FILTER (BOSTON IIIA)
OIL, 19 imp gal.
FLAP ACTUATOR
RETRACTION ACTUATOR
WRIGHT CYCLONE GR-2600, 1600 hp
STATIC
MAX.EXT.
0 1 2 3 4 5 6 7 8 9 10 ft.
0 1 2 3 m.
RUDDER HINGES
TAILPLANE ROOT SECTION
FLARES
TWIN .303 BROWNINGS
VICKERS .303 (DEPLOYED)
SPARE MAGAZINES
RADIO RACKS
HEATER UNIT
WING ROOT SECTION
LIFE RAFT
CANOPY (OPEN)
HYDRAULIC RESERVOIR
BOMB DOORS (OPEN)
BOMB DOOR ACTUATOR
FLARE TUBES
NOSE HATCH
*SCRAP VIEW: REAR TURRET, A-20G
MARTIN POWER TURRET
REVISED RADIO EQUIP.
FABRIC COVERED AILERON
OUTBOARD FUELLING POINT
51 imp. gal.
BOMB AIMER/OBSERVER STATION
110 imp gal.
GUN STOWAGE DOORS
ESCAPE HATCH
INBOARD FUELLING POINT
SLIDING GUN HATCH
FABRIC COVERED ELEVATOR
OUTER WING JOINT COVER
2× .50 cal.
INCREASE IN WIDTH
*SCRAP VIEW: NOSE SECTION, A-20G
*SCRAP VIEW: MARTIN UPPER TURRET, A-20G

(nose, ventral and dorsal) each housing a pair of 0.30in machine guns. This was a one-off experiment first flown in mid 1941.

By the time the A-20C Havoc appeared in early 1941, a high degree of standardisation between American and British versions had been achieved with only minor diversions in equipment. The A-20C was equivalent in almost all respects to the RAF's Boston III and compared to the A-20B again featured self sealing fuel tanks and increased armour protection. Gun armament reverted to 0.30in machine guns in all positions (the unsuccessful nacelle guns were deleted) and the diagonal side nose transparency design was reintroduced. The A-20C was significant in that it was the first DB-7 variant capable of carrying a 2,000lb (907kg) standard naval torpedo. A-20C production amounted to 948 aircraft, 808 by Douglas at Santa Monica and 140 by Boeing.

The 1943 conversion of 13 A-20Cs to night fighters resulted in the P-70A, fitted with radar and six 0.50in machine guns in a solid nose. There was no A-20D, the A-20E designation was applied to 17 A-20As fitted with R-2600-11 engines for performance testing and the XA-20F was a single experimental conversion of an A-20A fitted with a 37mm cannon in the nose and power turrets in the dorsal and ventral rear fuselage positions.

The Peak of Development

The most numerically and operationally important version of the Havoc was the A-20G, no fewer than 2,850 of them rolling off the Santa Monica production line from 1942. The A-20G was modified considerably during its production life and this was the first bomber version to feature a solid nose with a heavy fixed gun armament. All 'G' models were powered by 1,600hp R-2600-23 engines and maximum weight grew from 22,000lb (9,979kg) to 27,200lb (12,338kg) and even to 30,000lb (13,608kg) for ferrying purposes as the war progressed. These increases degraded the maximum speed slightly (to about 339mph) but the new features more than compensated.

As originally produced, the A-20G featured four 20mm cannon and two 0.50in machine guns installed in the nose along with a single 50 calibre in the usual dorsal flexible mount. The ventral flexible gun was also changed to a 0.50in unit. Bomb load remained the previous 2,000lb (907kg) but the fuel capacity was increased to 445imp gal (2,023 l) by the fitting of extra tanks in the bomb bay.

The cannon were found to be prone to jamming and were replaced in subsequent production blocks by four 0.50in machine guns, giving a total of six in the nose. Most of the cannon equipped A-20Gs were despatched to the Soviet Union.

The biggest single advance in the Havoc series came with aircraft from A-20G Production Block 20 and subsequent. These aircraft were fitted with a dorsal Martin power turret containing two 50 calibre guns and in order to accommodate the turret the rear fuselage was widened, allowing a further increase in fuel capacity to 605imp gal (2,750 l). With an additional under fuselage drop tank of 310imp gal (1,409 l) capacity fitted, the A-20G was capable of being ferried just over 2,000 miles (3,200km). Other improvements were added to the A-20G as its production progressed including more substantial

DOUGLAS DB-7B BOSTON III
Powerplants: *Two Wright R-2600-A5B Double Cyclone 14 cylinder, two row, air cooled radial engines of 2,603cu in (42.65 litres) capacity and fitted with two speed supercharger. Rated at 1,600hp for takeoff and at 1,000ft and 1,400hp at 10,000ft. Hamilton Standard Hydromatic three bladed constant speed fully feathering propellers of 11ft 3in (3.43m) diameter. Fuel capacity 329imp gal (1,495 l) in four self sealing tanks in the inner wing panels, one on each side of the engine nacelles.*
Dimensions: *Wing span 61ft 4in (18.69m); length 47ft 6in (14.48m); height 17ft 7in (5.36m); wing area 465sq ft (43.2sq m).*
Weights: *Normal max takeoff 20,320lb (9,217kg); overload 23,500lb (10,660kg); typical empty 15,051lb (6,827kg); loadings (normal max t/o weight) power 6.35lb/hp, wing 43.7lb/sq ft.*
Armament: *(Bombs) maximum load of 2,000lb (907kg) in split bomb bay comprising four 500lb (227kg) bombs; normal load 1,000lb (454kg) comprising two 500lb (227kg) or four 250lb (113kg) bombs.*
(Guns) four fixed Browning 0.303in machine guns in nose with 500 rounds per gun; twin Browning 0.303in machine guns in dorsal position on flexible mount; one Vickers K 0.303in machine gun in ventral position on flexible mount.
Performance: *Max speed 311mph (500km/h) at sea level, 338mph at 12,500ft; cruising speed 273mph (439km/h); initial climb 2,000ft (609m)/min; service ceiling 27,600ft (8,412m); range with 2,000lb (907kg bomb load 525 miles (844km); range with 1,000lb (454kg) load 745 miles (1,199km).*

Two-thirds of the 995 A-20B Havocs built in 1941 were delivered to the Soviet Union. This shot was taken at a staging airfield in Iran where A-20Bs, T-6s and P-40s await pilots.

DOUGLAS A-20G HAVOC
Powerplants: *Two Wright R-2600-23 Double Cyclone 14 cylinder, two row, air cooled radial engines of 2,603cu in (42.65 litres) capacity fitted with two speed superchargers. Rated at 1,600hp for takeoff; 1,675hp war emergency; 1,400hp at 10,000ft. Hamilton Standard Hydromatic three bladed constant speed and feathering propellers of 11ft 3in (3.43m) diameter. Normal fuel capacity 329imp gal (1,495 l) in four self sealing wing tanks; extra 270imp gal (1,227 l) in bomb bay tanks; underfuselage 310imp gal (1,409 l) drop tank for ferrying only.*
Dimensions: *Wing span 61ft 4in (18.69m); length 48ft 0in (14.63m); height 17ft 7in (5.36m); wing area 465sq ft (43.2sq m).*
Weights: *Normal combat takeoff 24,000lb (10,886kg); max combat 27,200lb (12,338kg); max ferry overload 30,000lb (13,608kg); basic empty 17,200lb (7,802kg); loadings (at 27,200lb) power 8.5lb/hp, wing 58.5lb/sq ft.*
Armament: *(Bombs) maximum load of 2,000lb (907kg) in split bomb bay plus (later models) 2,000lb (907kg) on four underwing hardpoints.*
(Guns) Six fixed Colt-Browning 0.50in machine guns in nose with 350rpg; two 0.50in machine guns in powered dorsal turret (later models) with 400rpg; one 0.50in machine gun on ventral flexible mount with 400 rounds. Earlier models had two 0.50in dorsal machine guns on flexible mount.
Performance: *Max speed 339mph (545km/h) at 12,400ft, 317mph (510km/h) at 10,000ft; cruising speed 230-272mph (370-438km/h); initial climb 1,300ft (396m)/min, time to 10,000ft 7.1min; service ceiling 25,800ft (7,864m); range (609imp gal, 2,000lb bombs) 1,025 miles (1,649km) at 238mph (383km/h), max ferry range 2,035 miles (3,275km).*

The most numerous Havoc variant was the A-20G, introduced in 1942 and substantially developed up until the end of the war. Note the power operated turret in the 'mid upper' position.

armour and better bomb sights and navigational equipment. In addition, the general structural strengthening associated with the later A-20Gs (from Block 20 onwards) allowed a doubling of the ordnance load to 4,000lb (1,814kg) with the fitting of underwing racks outboard of the engines, each with a capacity of 500lb (227kg).

Some A-20Gs were converted to P-70 night fighter configuration. Twenty-six became P-70A-2s with the standard G nose armament while a single example became the P-70B-1 with radar and radio in its solid nose and three 0.50in machine guns on each side of the forward fuselage mounted in blisters. A further 105 Havocs (A-20Gs and Js) were converted to similar standards as P-70B-2s.

The A-20J Havoc with frameless transparent 'bombardier' nose came about to meet a requirement for bombing formation 'lead ships' with full navigation facilities. A total of 450 A-20Js was built in 1943, of which 165 went to the RAF as Boston IVs. Apart from the nose design and its associated armament (reduced to a pair of 'fifties') the A-20J was otherwise similar to the A-20G, including its powerplant.

The A-20H and A-20K were solid and glass nosed, respectively, developments of the A-20G/J which differed from their predecessors in having 1,700hp R-2600-29 engines. Production amounted to 412 A-20Hs and 413 A-20Ks in 1944, 90 of the latter being supplied to the RAF as the Boston V.

Head on shot of a USAAF A-20G showing the nose armament which initially was four 20mm cannon and two 0.50in machine guns. This was soon changed to six 'fifties'. (RAAF)

EARLY DAYS IN AUSTRALIA

The Douglas Boston was introduced into Royal Australian Air Force service as much by accident as design but nevertheless went on to perform a particularly important role in the New Guinea area in the hands of the RAAF's No 22 Squadron. This was the only RAAF squadron to operate the type, finally relinquishing them for Beaufighters at the end of 1944 after more than two years' active service in New Guinea. A total of 69 of the A-20/Boston Havoc series was taken on RAAF charge, with serial numbers in the A28 series.

The 69 RAAF Bostons comprised four variants: 11 DB-7B Boston IIIs and 11 Boeing built Boston IIIAs (A28-1 to -22) delivered in March and April 1942; nine A-20Cs (A28-23 to -31) delivered between September and November 1943; nine A-20As (A28-32 to -40) taken on charge in November 1943 and 29 A-20G Havocs (A28-50 to -78) delivered between June and October 1944.

Apart from the installation of British guns and some equipment, the A-20C was to all intents and purposes the same aircraft as the DB-7B and was in fact called the Boston III by the British. The A-20Cs delivered to Australia retained the standard American armament when delivered.

Refugee Bostons

The first Batch of 22 DB-7B Boston III/IIIAs to enter RAAF service was intended for the Netherlands East Indies Air Force (now Indonesia) and was on the water when the Japanese overran that part of the world. Instead they (and other aircraft cargoes intended for the NEI) were diverted to Australia as "Refugee Cargo". They had come from Royal Air Force stocks and carried British serial numbers, markings and standard RAF dark earth/dark green camouflage with 'sky' undersurfaces.

According to research carried out by 22 Squadron historian Dick Hutchinson, by the time the Bostons arrived in Australia during March and April 1942 "the 22 DB-7Bs were still sporting the basic RAF scheme but had had the British markings overpainted with those of the *Kon Marine* (Dutch Navy). The tail flashes had been painted out with the basic camouflage colour corresponding to the relevant pattern. The roundels had been overpainted with the Dutch black bordered triangle and some evidence suggests that these markings were also added underwing. The RAF serial number was overpainted in the background colour before being replaced with the three number Dutch serial and the words 'KON MARINE' appeared in small black letters on either side of the nose".

Dick Hutchinson also suggests that the nose 'KON MARINE' markings were not always overpainted when the Bostons were taken on charge by No 22 Squadron RAAF and some aircraft retained this feature even when flying operations in New Guinea later in 1942.

The 22 DB-7Bs arrived in Australia in five batches during March and April 1942, their journey by sea terminating in Melbourne on board the following ships: SS *Tossari* (3 March, three aircraft), SS *Tarakan* (6 March, five aircraft), SS *Mapia* (3 April, six aircraft), SS *Tabinta* (7 April, three aircraft) and SS *Weltevroden* (26 April, five aircraft).

From there, the Bostons were shipped in their crates for assembly at RAAF Laverton, a process which resulted in some abnormal problems for those involved as instruments, notices and notes were in Dutch and had to be translated, rewritten and where appropriate remanufactured.

Into RAAF Colours

The Bostons were allocated to the RAAF's No 22 Squadron upon their arrival in Australia, but before the squadron could fully settle down with its new aircraft, one or two 'administrative' matters had to be sorted out.

These revolved around the formation of No 18 (Netherlands East Indies) Squadron at Canberra in early April 1942 and intended to utilise 'refugee' aircraft and Dutch personnel displaced by the Japanese invasion of the NEI. The squadron would operate under the control of the RAAF (albeit as part of the Netherlands East Indies Air Force from June 1942) and would feature a high pro-

An oft-used official photograph of one of the first batch of RAAF DB-7B Boston IIIs. The first three aircraft arrived in Melbourne on board the SS Tossari on 3 March 1942. (RAAF PR)

The caption attached to this photograph describes this Boston as the "first in Australia" but fails to identify its serial number or location. The first eight Bostons were officially taken on RAAF charge on 29 March 1942. (via Mike Kerr)

portion of Australian air and ground crew working alongside their Dutch counterparts.

18 Squadron's initial equipment comprised five B-25 Mitchells and 15 Bostons, ten of which were from the batch of 22 DB-7Bs and five of which were A-20As from USAAF stocks. The ten DB-7Bs which served with No 18 Squadron were the RAAF's A28-4, 5, 8, 9, 11, 12, 15, 16, 18 and 22. The others remained with No 22 Squadron, which had received its first Bostons at its Richmond NSW base on 26 April 1942.

RAAF BOSTONS
SUMMARY OF DELIVERIES

RAAF Nos	Model	Qty	Delivery
A28-1/22	DB-7B	22	03-04/42
A28-23/31	A-20C	9	09-11/43
A28-32/40	A-20A	9	11/43
A28-50/78	A-20G	29	06-10/44

Notes: Of the 22 DB-7B Boston IIIs (A28-1/22), 11 were Boeing built Boston IIIAs. These are A28-2, 10, 11, 12, 13, 15, 16, 17, 18, 19 and 20. Serial numbers A28-41 to -49 were not allocated.

The ten mentioned above would only temporarily appear on the strength of No 18 NEI Squadron before returning to No 22 Squadron, although this was not the original intention, and the Boston served with the squadron as a major component of its strength for only 12 days from the middle of June 1942.

On the 13th of that month, a group of USAAF pilots arrived in Canberra to take 18 Squadron's five Mitchells away. The squadron's commanding officer, Lt Col B J Fiedeldij, refused permission for them to do so, even (allegedly) going as far as having some of them locked up. There were several reasons for this altercation, among them the fact that the Dutch preferred the Mitchell over the Boston. It was claimed that 18 Mitchells had in fact been promised. The stalemate was broken the following day and 18 Squadron was allowed to keep its Mitchells and would in fact standardise on the type. The 18 Squadron story is discussed in greater detail in the 'Mitchell' section of this book.

On 23 June the RAAF's No 22 Squadron got its Bostons back, while other aircraft (A-20As) which had also appeared on the strength of the Dutch squadron would eventually find their way into RAAF service. Of these, four (A28-32, 33, 34 and 36) later served with No 22 Squadron.

22 Squadron in Australia

Immediately prior to receiving its Bostons, No 22 Squadron had flown Wirraway single engined general purpose trainers from its Richmond base, its primary role being to provide a modicum of

Another of the first batch of Bostons shortly after delivery. On this example, the words KON MARINE can be seen painted on the nose. The photograph was most likely taken at Moruya NSW during 1942. (22 Squadron Historical)

Drawing of a Boston III cockpit with a photograph of an A-20G cockpit for comparison. KEY TO DRAWING: 1. tank crossfeed control; 2. engine crossfeed control; 3. LH and RH fuel tank selectors; 4. carburettor air temperature controls; 5. supercharger controls; 6. fuel valve panel light; 7. prop control friction nut; 8. engine control friction nut; 9. mixture controls; 10. throttle controls; 11. propeller controls; 12. lower electrical panel; 13. engine controls light; 14. upper electrical panel; 15. ignition switch; 16. control column and wheel; 17. control lock release; 18. fixed gun firing switch; 19. engine primer pump; 20. manifold pressure gauge drain cock; 21. radio controls light.
22. recognition lights Morse key; 23. oxygen outlet; 24. telegraph tapping key; 25. recognition lights switches; 26. R-3003 receiver master switch; 27. radio control panel; 28. compass light; 29. trim tab controls light; 30. cockpit extension light; 31. rudder trim control; 32. elevator trim control; 33. aileron trim control; 34. landing gear emergency release; 35. landing flare release; 36. fuel dump valve control; 37. compass; 38. bomb control panel; 39. oxygen regulator; 40. heating/ventilation outlet control.
41. fixed gun charging controls; 42. rudder and brake pedals; 43. rudder pedal adjustment release; 44. wobble pump; 45. bomb door control; 46. oxygen regulator; 47. heating/ventilation controls; 48. upper cowl flap control; 49. safety harness release; 50. lower cowl flap/oil cooler controls; 51. seat adjustment; 52. hydraulic hand pump by-pass valve control; 53. wings flaps control; 54. landing gear hydraulic control; 55. hydraulic hand pump control.
56. control lock; 57. parking brake; 58. turn and bank indicator valve control; 59. outside air temperature; 60. clock; 61. suction gauge; 62. steering indicator; 63. airspeed indicator; 64. altimeter; 65. landing gear/wing flaps position indicator; 66. gyro horizon; 67. directional gyro; 68. rate of climb indicator; 69. turn and bank indicator; 70. LH manifold pressure gauge; 71. RH manifold pressure gauge.
72. LH tachometer; 73. RH tachometer; 74. LH oil pressure gauge; 75. RH oil pressure gauge; 76. LH fuel pressure gauge; 77. RH fuel pressure gauge; 78. hydraulic pressure gauge; 79. LH oil temperature gauge; 80. RH oil temperature gauge; 81. LH carburettor air temperature gauge; 82. RH carburettor air temperature gauge; 83. LH cylinder head temperature gauge; 84. RH cylinder head temperature gauge; 85. fuel quantity; 86. armour plate release.

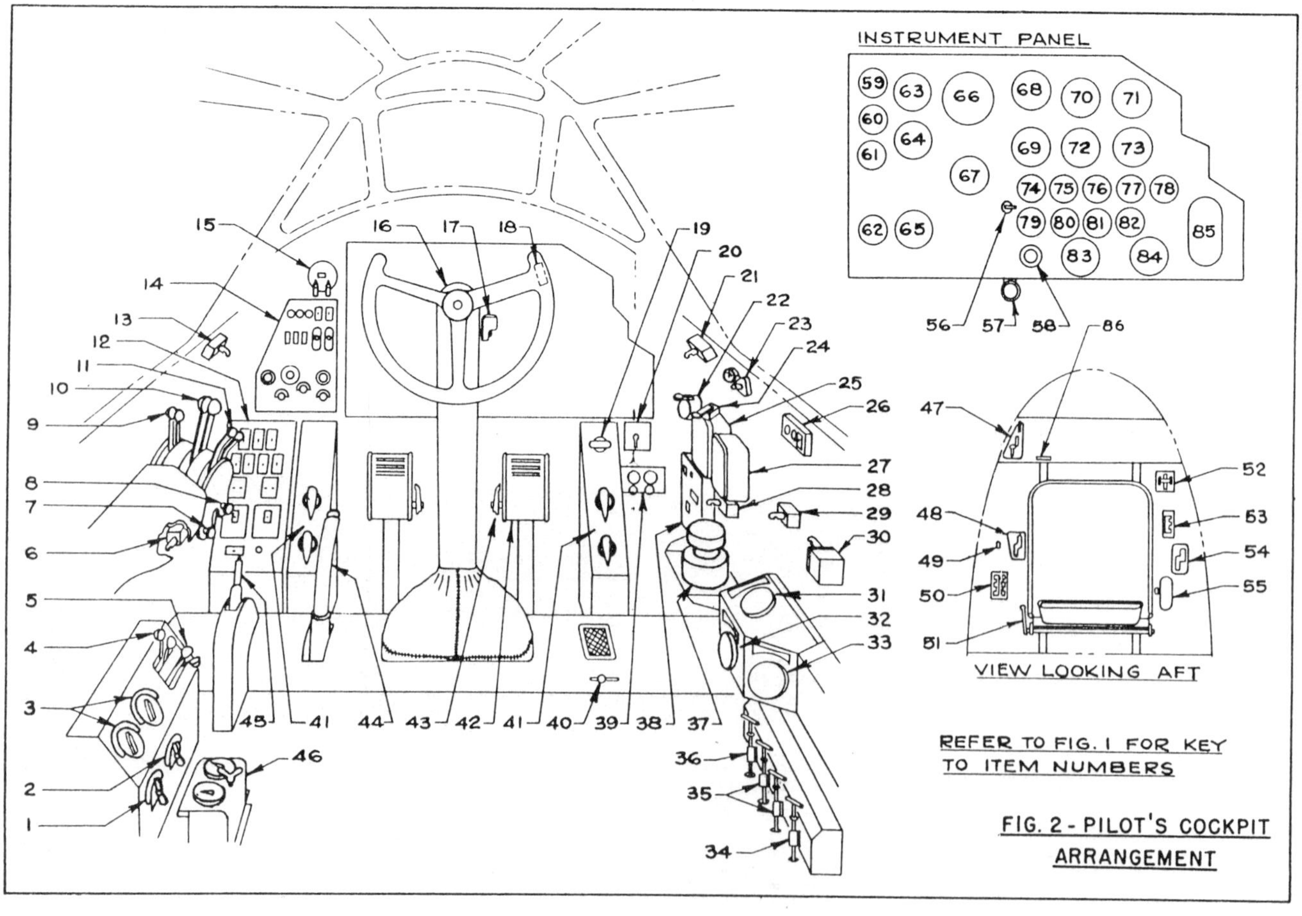

Before being sent to New Guinea in September 1942, No 22 Squadron's Bostons were modified with the glazed nose panels replaced with painted aluminium and four 50-calibre machine guns installed. The navigator usually occupied the nose of the aircraft but in 22 Squadron's Bostons he was deleted and the crew reduced from four to three. (22 Squadron Historical)

air defence capability for the city of Sydney. The squadron had been formed as a Cadre unit at Richmond in April 1936, equipped with two Hawker Demons and three de Havilland Gipsy Moths. Avro Ansons were added in 1937 and Wirraways in October 1939.

Conversion courses onto the Boston began with the arrival of the first aircraft in April 1942 and operations with the new attack bomber began the following June, by which time the last Wirraways had been allocated to other units. The squadron's commanding officer at the time was Flt Lt R E Bell and its primary activities whilst based at Richmond were training and anti-submarine patrols.

The latter activity produced some action for the squadron's pilots as Japanese submarine activity off the coast of New South Wales was considerable and had been responsible for the loss of a substantial amount of shipping.

Three attacks on suspect submarines were recorded by 22 Squadron in mid 1942. The first occurred on 6 June when A28-12 flown by Flt Lt Morgan was credited with the destruction of a submarine just off Sydney's Botany Bay. Another attack was made three days later when Flg Off Miles and his crew claimed to have sunk a submarine off Williamtown, near Newcastle. The third attack, in August, involved Sgt Smith and A28-18 off Tuggerah, north of Sydney. A claim for a damaged submarine was made after three 250lb bombs were dropped near the target, but this and the two sinking claims were dismissed after the war following further investigation.

No 22 Squadron's time at Richmond was not without cost, two aircraft being written off and another damaged, two of them due to the same problem. The first casualty was A28-17 which was damaged beyond repair due to an engine fire whilst running up at Richmond on 18 May and A28-2 was suffered a similar fate the following day when it failed to become airborne during takeoff at Richmond.

The aircraft ploughed through fences and other obstacles before coming to

Boston A28-19 shortly after its arrival in Australia, displaying its RAF camouflage scheme. The rectangular black patch just above the nose gun blister covers the words KON MARINE. Note the aircraft's perspex nose, the form in which these Bostons were delivered. (22 Squadron Historical)

Local modification: a fixed 0.30in machine gun was installed in the tails of a couple of Bostons in New Guinea. The gun was fired electrically by the rear gunner and was fixed. It was intended to scare off enemy aircraft attacking from the rear. (22 Squadron Historical)

rest, Plt Off R Fethers and his crew escaping unhurt. The pilot reported that the Boston's control column would not come back, but investigation revealed he had simply not pulled hard enough, the column requiring a considerably hefty pull to get the nosewheel off the ground and the aircraft into flying attitude during the takeoff roll. Another Boston was bent at Mascot (Sydney) airport under similar circumstances.

Preparations For New Guinea

From the time it was re-equipped with Bostons, No 22 Squadron had been expecting an overseas posting to active service in the New Guinea area. At the beginning of September 1942 the squadron was informed it would be moving to that region. At the time, the Japanese were established in many parts of New Guinea and were continuing in their efforts to take the strategically vital town of Port Moresby. They had received a major setback at Milne Bay where Australian forces had inflicted their first major defeat on land, thus denying them half of the planned two-pronged thrust on Port Moresby. Now the Allies' objective was to push the Japanese back.

An interesting shot of one of the nine A-20Cs supplied in late 1943. Note the aluminium nose panels, the three machine guns and the camera port. (22 Squadron Historical)

A poor quality shot but an interesting one of A-20A Boston (A28-39) in natural metal finish. This aircraft served with No 22 Squadron between March and July 1944 whereupon it returned to Australia where No 3 Aircraft Depot at Amberley removed its armament and armour plating. It was then used for pilot conversion training and for the training of test pilots. In 1946 it became an instructional airframe for the Queensland Air Training Corps. (via Neil Mackenzie)

A 22 Squadron advance party left for their new base at Ward's Strip, Port Moresby, on 4 September followed by 14 Bostons six days later and the main body of the squadron by sea from Townsville on the 14th. This group had travelled by road from Richmond to Townsville in north Queensland. The aircraft travelled from Richmond via Townsville and arrived in Port Moresby on 19 October. They were stationed at Konedobu for a fortnight before moving on to Ward's Strip. One Boston (A28-19) was lost during the stopover at Townsville, crashing near Woodford Qld on 9 October.

Before departing for New Guinea, the Bostons received some modifications, notably to their gun armament. As DB-7B models, they all featured glazed noses but these were replaced with painted aluminium panels. Inside the nose, the structure was modified to house four forward firing 0.50in Browning machine guns, which in combination with the standard two pairs of '303s' mounted in blisters on the side of the nose, gave the Bostons formidable hitting power when used in the ground attack/strike role.

For operations in New Guinea the standard crew of four (pilot, navigator, and two gunners was reduced to three, with the navigator (who usually occupied the glazed nose section) being omitted.

With the arrival of 22 Squadron's Bostons in New Guinea, a notable chapter in the Royal Australian Air Force's history was about to begin.

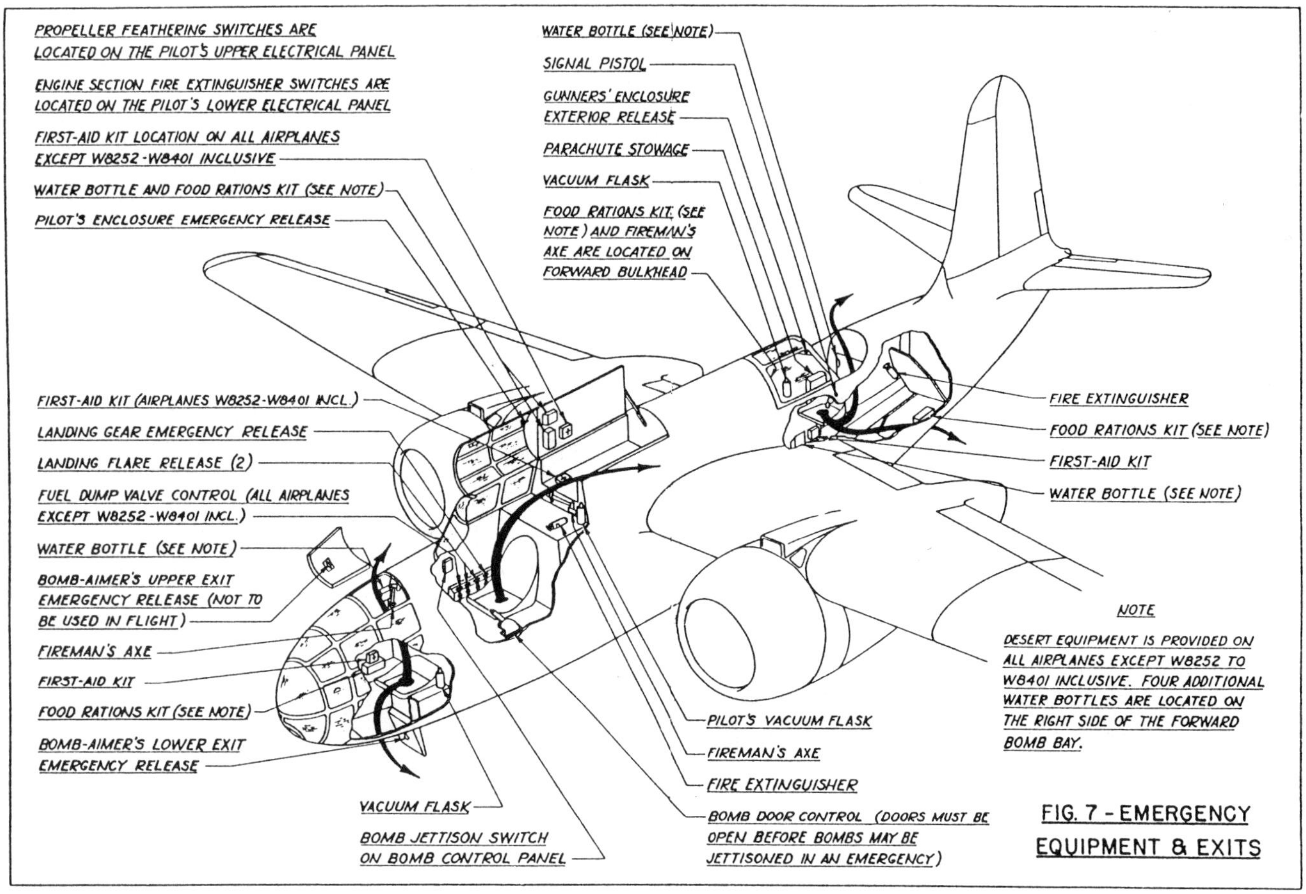

FIG. 7 - EMERGENCY EQUIPMENT & EXITS

A28-30, an A-20C Boston in 22 Squadron markings. Note the three gun nose, the flat camera port immediately below them and the lack of 'cheek' guns on the side of the nose. (22 Squadron Historical)

BOSTON TEA PARTIES – 22 SQUADRON AT WAR

It was perhaps inevitable that No 22 Squadron's operations in New Guinea should be given the good-natured nickname 'Boston Tea Parties', but in reality the squadron's personnel and aircraft endured conditions which were far removed from anything resembling the niceties and refinement of a real tea party.

Operating from airstrips cut from the jungle, flying in often atrocious and always dangerous weather and geographical conditions and enduring living conditions which could best be described as primitive was a severe test of not just 22 Squadron but of any fighting unit in New Guinea.

Monsoonal rain one minute, stifling heat and humidity the next, dust or mud in copious quantities, the necessity for aircraft to be maintained, fuelled and armed out of doors, a resolute enemy, little chance of rescue or survival if forced down ... paradise lost.

First Operations and Problems

No 22 Squadron's Bostons went into action for the first time on 15 November 1942 when three aircraft (led by the squadron's CO, Flt Lt R Bell) carried out an armed reconnaissance along the coast from Sananda Point to the mouth of the Kumuai River. Most of the early missions were of a similar nature but included a number of bombing and strafing attacks against Japanese troops, land and sea communications and installations in the Buna area on the northern coastline of New Guinea's eastern peninsula. At this stage Australian troops were advancing on Buna after their defeat of the Japanese on the Kokoda trail.

The Bostons sometimes operated alongside No 30 Squadron's Beaufighters from Port Moresby, the latter type becoming something of a specialist in low level strafing missions and proving superb in that role against Japanese barges, troop and gun positions, supply dumps, vehicles and ships. The Bostons

Two shots of a 22 Squadron formation over the Bismark Sea area in November 1942. The aircraft are A28-16 'R', A28-8 'J', A28-15 'Q' and A28-5 'F'. None survived beyond September 1943 although the remains of A28-8 have been recovered for restoration by the RAAF Museum (22 Squadron Historical/via Mike Kerr)

attacked similar targets but used bombs as their primary armament with the battery of machine guns grouped in and around the nose providing a more than useful secondary strafing capability.

The loss of three Bostons under similar circumstances took much of the gloss off the squadron's first month of operations. On 11 November, A28-12 (pilot Flt Lt V Morgan) blew up during bombing practice on a wreck off Port Moresby; on 26 November A28-22 (Flt Lt K McDonald) blew up whilst attacking Buna Strip; and on 29 November A28-20 (Flt Lt H Bullmore) suffered a similar fate whilst attacking Gona Mission.

Investigation revealed a common factor in these accidents – the carrying of 20lb fragmentation bombs in the Bostons. The lightness of the bombs meant that when released into the slipstream in a cluster, one would blow back against another and cause a concussive chain reaction, resulting in a large explosion immediately below the aircraft, with disastrous results.

There were no further such incidents when use of the fragmentation bombs was stopped, although 20lb bombs of a different type were subsequently used by the squadron with no problems. As part of the investigation, a Boston dropping a load of defused 20lb fragmentation bombs was filmed by famed Australian war photographer, Damien Parer. After 22 Squadron advised RAAF Headquarters in Melbourne of the results of the investigation, Headquarters then advised the squadron that they already knew that Bostons dropping similar loads in North Africa were suffering the same problem!

A remarkable photograph showing the destruction of Boston A28-12 (flown by Flt Lt V Morgan) over Port Moresby Harbour in November 1942 during bombing practice. The aircraft's 20lb fragmentation bombs exploded immediately after leaving the aircraft, with obvious results. Two other Bostons were lost for the same reason during the month before an investigation uncovered the problem and the bombs were withdrawn from use. (22 Squadron Historical)

A28-4 at Ward's Strip in January 1943 following its crash landing after being damaged in action. Flt Lt Wines was the pilot. (22 Squadron Historical)

Boston A28-11 DU-M over the Owen Stanley Ranges in New Guinea. This aircraft survived until early 1945 when it was converted to components. (22 Squadron Historical)

On the subject of armament, two Bostons (A28-8 and 9) were fitted with a 0.303in machine gun in the extreme tail. Electrically operated and fired by the mid-upper gunner, the guns couldn't be properly aimed and were used to scare off enemy fighters attacking from the rear. A couple of Bostons also had a single 0.303in Vickers K machine gun mounted in the rear floor of the aircraft and rearward firing guns were also installed in each engine nacelle on a small number of aircraft. The latter were operated by pulley and wire by the rear gunner but were unsuccessful and quickly removed, as were the floor guns.

December 1942 opened with three Bostons attacking a Japanese convoy comprising four destroyers which were attempting to reinforce Buna, while on the 4th of the month, one of the many interesting incidents of 22 Squadron's time on Bostons took place.

On that day, five Bostons bombed the Salamaua area, and whilst returning from that strike, the aircraft of Flg Off G Turner encountered bad weather over the Owen Stanley mountain range and was forced to drop out of the formation. The Boston disappeared from sight and failed to return to base. A search was carried out the next day and located the aircraft, apparently intact, in a small clearing near Wairope near Kokoda. It

(left) Ward's Strip Port Moresby under Japanese attack. 22 Squadron's Bostons survived the attentions of the Japanese until 1945 when a raid on Morotai all but wiped out the squadron. (22 Squadron Historical)

(right) 'Indoor Sport!!' was Boston A28-16/ DU-R, the aircraft of Dick Fethers. Both were lost in action over Gasmata in September 1943.

was decided to send a salvage party of four from the squadron which included Flt Lt Claude Sladen.

The party flew to Kokoda and then went on foot to the aircraft, a journey of about 25 miles (40 kilometres) through thick jungle. Carrying supplies and tools, the party took two days to reach the Boston and on arrival began stripping it of all excess weight, including guns, ammunition, equipment and all but that fuel which was considered necessary to fly back to Port Moresby. Several days work resulted in the aircraft being prepared to fly and a strip being cut in the jungle which was calculated to be just long enough to allow the Boston to takeoff.

Another few days' wait for the weather to clear and the sodden and muddy ground to dry off a little followed and then Sladen – flying solo – flew out of the extended clearing "with inches to spare" and took the Boston back to Moresby in an example of superb airmanship. The whole exercise was something of a triumph of determination and ingenuity, although three of the four involved in it – Flg Off Chester, Sgt B Williams and Sgt G Kendall – still had some work to do after Sladen took off. They had to walk back through the jungle to Kokoda, in company with Flg Off Turner and his crew.

Intense Activity

Now under the command of Sqdn Ldr Keith Hampshire, 22 Squadron continued an ever increasing sortie rate in December 1942 and into the New Year. Convoys, troop emplacements, airfields, stores depots and just about anything controlled by the Japanese were attacked by the Bostons and armed reconnaissance sweeps were carried out along the north-east coast of New Guinea along with low level attacks on Lae.

By now the Allies were gaining the upper hand in the New Guinea campaign with Buna and Sananada Point falling to the Australians and Americans in January 1943. That month was the busiest of the squadron's first year in the area: 204 hours flown on operations, 80 250lb and 80 500lb bombs dropped and 83,400 rounds of ammunition expended.

Despite its setbacks and two unsuccessful attempts to take Port Moresby, Japan did not give up in New Guinea. In January 1943 another offensive was begun, this time in the Wau area. By the end of the month the campaign was being repelled and Japanese forces were being forced back with 22 Squadron's Bostons among those which bombed, strafed and generally harassed the Japanese and their supply line as they fell back during February.

No 22 Squadron personalities 'borrow' 30 (Beaufighter) Squadron's mascot. From the left: Sladen, Rowell, Fethers, McMasters. The animal was orphaned by bombing and adopted by the squadron. (22 Squadron Historical)

The squadron flew twelve operations in February 1943, eight of them in the Mubo and Komiatum areas against villages in front of the advancing Australian troops. These raids were extremely effective with hundreds of Japanese casualties being recorded and an easier path with less resistance being provided for the troops. Lae was another target for the Bostons during the month.

This period of intense activity inevitably resulted in more losses for the squadron. In December 1942 A28-1 had crashed on takeoff at Kiriwina and was reduced to spares, while in January, A28-4 crash landed at Jackson Strip, Port Moresby, after being badly damaged by anti-aircraft fire during a raid and was scrapped. Two more were lost in February 1943: A28-14 failed to return from a raid on supply dumps at Malahang after disappearing from the returning formation in bad weather, and A28-21 which was seen burning and crashing into the sea near Salamaua, probably as a result of anti-aircraft fire.

By March 1943 then, nine of the original 22 Bostons had been lost, including those casualties which had occurred before 22 Squadron moved to New Guinea. Another four aircraft would be lost before replacement Bostons began arriving the following September.

On 3 March 1943 some of 22 Squadron's Bostons were among the Allied aircraft called in to attack a Japanese convoy in what became known as the Battle of the Bismark Sea. This photograph, taken during the action, shows the Japanese destroyer successfully attacked by the Bostons. (22 Squadron Historical)

The Battle of the Bismark Sea

Despite earlier setbacks, Japan had still not abandoned its plans to take Port Moresby by March 1943, and during the first week of that month the decisive Battle of the Bismark Sea took place. This was an operation mounted against a large Japanese convoy which had been assembled to reinforce Lae. Conducted over three days (March 2-4), the battle resulted in the sinking of all eight Japanese transport ships and four of the eight escorting destroyers for the loss of six Allied aircraft. Both Australian and American aircraft were involved and the battle contributed greatly to the end of Japanese aspirations in New Guinea.

No 22 Squadron's contribution to the victory was substantial, with two attacks on enemy held Lae airfield and participation in a co-ordinated attack by more than 100 Allied aircraft on the convoy during the afternoon of 3 March. The attacks on Lae airfield were designed to subdue Japanese fighter activity and their support of the convoy.

Five Bostons, led by Sqdn Ldr C C (Charles) Learmonth, participated in the attack and despite interception by enemy fighters, all returned safely. On that day, 22 Squadron's Bostons flew a total of 20 sorties without loss.

The following is a report on the convoy action by Sqd Ldr Learmonth, who had been appointed Temporary Commanding Officer following Keith Hampshire's wounding by shrapnel during one of the mopping up operations after the battle ...

ATTACK ON JAPANESE CONVOY – 3rd MARCH 1943

Report of attack carried out by Boston aircraft of No 22 Squadron RAAF

1. ORDERS: At 1300L on 3rd March 1943, the following orders were received from the Controller, No 9 OG: All available Bostons to carry out co-ordinated attack on enemy convoy last reported position 07.05 South, 148.30 East widely scattered at 1220/3 March. Aircraft rendezvous at Cape Ward Hunt at 4,000 feet.

Set course for rendezvous at 1400 hours l/3 March, remaining in visual contact during approach to target. Rally point Cape Warla. Primary targets transports, first alternative warships, second alternative Lae aerodrome.

Order of assembly and approach to target:- one squadron of B-17s at 9,000 feet, one squadron of B-25s at 8,000 feet, one squadron B-25s 7,000 feet, one squadron Beaufighters 6,000 feet, one squadron B-25s 5,500 feet, one squadron B-25C-1s 5,000 feet, one squadron A-20s 4,500 feet, one squadron Bostons 4,000 feet.

Order of attack:- first Beaufighters strafing, then B-25s, followed by A-20s and Bostons all masthead [height], then

B-17s from 7,000 to 10,000 feet, followed by one squadron B-25s 3,000 to 6,000 feet.

Bombs:- medium bombers 500lb demolition and instantaneous fuse, high bombers 1,000lb demolition instantaneous fuse, masthead attack 500 and 250lb five second delay. Fighter cover will have [radio] sets on 4895Kcs. Bomber flight leaders will identify and locate themselves relative to target when calling for fighters in any emergency. Enemy convoy protected by fighters.

2. BOMB LOAD: Each aircraft was bombed up with two 500lb instantaneous fusing, and two 250lb SAP one second delay.

3. BASE TO RENDEZVOUS: Five aircraft were available and all were airborne by 1330L, reaching the rendezvous at 1410L. The weather over the ranges was bad, and the squadron was fortunate in finding a break south of The Gap and got through at 14,500 feet. One squadron of B-25s reached rendezvous at the same time as ourselves, also one or two B-17s; but by approximately 1500L, all aircraft had assembled and at 1505L set course for convoy.

4. METHOD OF ATTACK:

(a) Order of Approach:- as the Beaufighters and A-20s had been unable to cross the ranges, the attack was led to the target in the following order: one squadron of B-17s, followed by three squadrons of B-25s, one squadron of Bostons and one squadron of B-17s. (This may be incorrect but is the general impression of all Boston pilots).

Flt Lt K McDonald and A28-22 'Retribution'. The combination was 22 Squadron's first operational casualty, blowing up over Buna when releasing 20lb fragmentation bombs over the target on 26 November 1942. (22 Squadron Historical)

Enemy shipping in New Britain receives the attention of 22 Squadron's Bostons. The squadron achieved considerable success in this role. (22 Squadron Historical)

(b) Constitution of Convoy:- the convoy was sighted from 4,000 feet through light cloud at 1530L in the approximate position of 07'10"S/147'40"E. It appeared to consist of three destroyers and four transports widely dispersed. All the transports were stationary and burning when sighted, and one destroyer was stationary and low in the water as if it had been hit previously. One other destroyer was in the vicinity of the transports and the remaining one was steaming north at approximately 20 knots.

(c) The Attack:- all aircraft in the formation carried out a low level attack from 4,000 to 1,500 feet and released their bombs in sticks of four. The target chosen was the stationary destroyer, which appeared on the left as we passed over the convoy. As the aircraft approached, they went into line astern and formed a circle around the target. Thus all aircraft were in a position to attack in quick succession and the destroyer's firepower was further confused.

For low dive attacks this method is highly commended, and it proved very effective on this occasion. One aircraft had to make a second run owing to interference by enemy fighters. Each stick of bombs straddled the target and two direct hits with 250lb bombs and approximately five near misses were scored.

The destroyer appeared to be of the *Fubuki* class and the attack on it lasted three to four minutes. After the bombing, one aircraft proceeded to a burning transport and strafed barges and troops in the water around it.

5. ENEMY OPPOSITION:

(a) Fighters:- several aircraft were attacked by Zekes before and during their bombing runs. These attacks were unsuccessful, partly due to the Bostons being strung out in line astern just prior to bombing, and they afforded each other a certain amount of protection.

(b) Anti-Aircraft:- the A/A opposition was light and ineffective and did not worry any of the aircraft.

(c) General:- the nature of the combined attack appeared to confuse all the enemy defences and effectively divided his firepower. The whole attack was carried out far more rapidly than was planned, and the enemy had to contend with all aircraft of all formations at the one time. The Bostons were almost bombing through strafing B-25s and the B-17s were bombing from medium height through both.

Enemy fighters did not attack the Bostons in force and single enemy aircraft did not press home their attacks with much tenacity. No casualties to personnel, and no damage were sustained.

6. COMBINED EFFORT: The destroyer in the vicinity of the transports was attacked by either B-17s or B-25s and suffered direct hits with heavy bombs, and blew up almost instantaneously. All ships received a thorough strafing from the B-25C-1s and the bombing from B-25s and B-17s appeared very accurate. B-25C-1s strafed the decks of the destroyer we attacked some 30 seconds before the leading aircraft made a run, and this must have helped a considerable amount in keeping the ship's guns from firing during our bombing.

The fact that the complete attack was effected almost simultaneously is again stressed. All pilots were convinced of the great assistance this afforded in dividing the enemy's defences, and although it must have caused a certain amount of risk, no damage resulted.

7. GENERAL: As the aircraft approached the convoy, seven to ten Zekes were seen apparently engaged in a sham fight several miles to the west. These tactics achieved no object, and the Zekes quickly came over and attacked the Bostons as previously stated.

On leaving the area it was apparent to all personnel that the convoy could not reform to make any semblance of an effective force. All ships excepting the one destroyer proceeding north were either sunk or sinking, and it is understood that this ship was sunk the following day.

Two days later – 5th March – a formation of four Bostons made a search for lifeboats and barges off Cape Ward Hunt. Several lifeboats and rafts were located but none contained live men, and no attacks were made.

ADDENDUM – Reason For Attack and Bomb Load Not Being as Ordered: On

22 Squadron personnel photographed at Port Moresby. The front row contains many of the squadron's well known pilots. From left to right (front row): Flg Off Harry Rowell, Flt Lt Harry Craig, Flt Lt Rocky Mullens, Flt Lt Bob Wines, Flt Lt Dick Hunt, Sqdn Ldr Charles Learmonth, Wng Cdr Keith Hampshire (CO), Flt Lt Claude Sladen, Flg Off Dick Fethers, Flg Off Austin Skinner, Flg Off Jack McMaster, Plt Off 'Buzz' Burn (Learmonth's wireless-air gunner). (RAAF Historical)

Another formation shot of 22 Squadron's Bostons over a typical New Guinea cloudscape taken in November 1942. A28-16 in the foreground. (22 Squadron Historical)

the morning of 3rd March all aircraft were placed on alert for possible attack against the convoy. No instructions were given as to bomb load and aircraft were bombed up with 2 x 500lb inst. GP and 2 x 250lb SAP one second delay. In view of the fact that orders were not received until 1300 hours and all aircraft were due at the rendezvous at 1400 hours, the bomb load could not be changed in the time available. Consequently, the attack took the form it did instead of a masthead attack as ordered.

Rounds fired against target: 300 x 0.5in, 1,490 x 0.303in.

Weather: low scattered cloud, base 2,000 feet, visibility 15 miles.

General: while over the enemy convoy several dogfights were observed between P-38s and Zeros in the course of which one Zero and one P-38 were seen falling away towards the water as though shot down.

★ ★ ★

As can be seen from the above report, the attack was not as co-ordinated as was planned, but as Squadron Leader Learmonth points out, this turned out to be to the Allies' advantage as it divided and confused the Japanese defences. Learmonth succeeded Sqdn Ldr Hampshire as 22 Squadron's commanding officer in June 1943.

A photograph taken from one of 22 Squadron's Bostons on 16 March 1943 during the raid on Salamaua, the action for which Newton was awarded the Victoria Cross. (22 Squadron Historical)

Newton, VC

No 22 spent the next four months supporting the work which was being done preparatory to the Allied assault on Lae and Salamaua. Still operating from

Wards Strip, the Bostons offered close support for the troops on the ground, striking at everything from barges to supply dumps, occupied villages and troop concentrations. Often operating with 30 Squadron's Beaufighters, the Bostons greatly contributed to the 'softening up' of the enemy, gradually removing pockets of resistance and elements of his vital supply line.

These operations were usually flown in the face of strong anti-aircraft fire and damage to the aircraft was often sustained. Two more Bostons were lost during this period, both of them on operations.

It was shortly after the Bismark Sea engagements in March 1943 that 22 Squadron recorded the actions that would eventually see the awarding of the RAAF's only Victoria Cross of the Pacific war.

Flt Lt William Ellis Newton was one of the many 22 Squadron pilots with reputations for bravery and tenacity in combat. On 16 March, Newton and his crew (Flt Sgt B Lyon and Sgt B G Eastwood) were flying A28-7 in company with six other Bostons on a raid on fuel dumps, stores and buildings on the foreshore of Salamaua. Despite intense anti-aircraft fire the raid was successful, with large fires being started.

Bill Newton was prominent in this action, leading the attack and choosing not to evade the ground fire but to press straight on to his target through it from half a mile out. A28-7 received four direct hits from anti-aircraft fire, taking out one engine and causing substantial damage to the wings, fuselage, systems, control surfaces and fuel tanks.

Regardless of all this, Newton pressed on and successfully delivered his bombs from low altitude, starting the fires which would eventually consume two 40,000 gallon (182,000 litre) fuel tanks along with the stores dumps and buildings. A28-7 was then flown the 180 miles (290km) back to Port Moresby over the Owen Stanley ranges, on one engine and with holes everywhere. Somehow, the Boston was repaired, returned to service and survived until June 1944 when it crashed on takeoff at Gurney Strip, Milne Bay.

This raid was probably typical of 22 Squadron's activities in New Guinea, at least in the sense of the danger facing the crews. Whilst Newton's Boston sustained considerable damage, two others were also hit by anti-aircraft fire but all got home. The participants in the raid represented the highest levels of skill in that kind of mission at the time – Newton, Hampshire, Fethers, Dillon, Sladen and McKay. Two days later, on 18 March, Newton and his crew flew A28-3 on another mission to Salamaua, this time against a single building which had been previously damaged. The sortie was Newton's 52nd, and the target was adjacent to anti-aircraft guns, making the exercise doubly difficult and hazardous.

Flt Lt W E Newton, the RAAF's only Victoria Cross winner of the Pacific war. Newton flew 52 missions with No 22 Squadron and it was his exploits of 16 March 1943 in a raid against Salamaua which earned him the award. Two days later, and against the same target, Newton's aircraft was shot down. He and one of crewmen survived but were captured and brutally executed by the Japanese.

Bombing from low level, Newton scored a direct hit on the building but just as the bombs struck, so did the effects of the anti-aircraft fire. The Boston burst into flames, but Newton managed to control it sufficiently to turn around and fly along the shoreline, the intention being to get as far away from the enemy positions as possible so as to give some chance of escape following the inevitable forced landing.

The Boston was brought down on the water just off shore. Newton and Lyon escaped from the aircraft before it sank but Eastwood did not. The two surviving airmen swam ashore but evaded capture for only one day.

Then followed acts of barbarism which were unfortunately typical of the Japanese and contrary to the rules of war. After interrogation, both airmen were executed, Lyon by bayonet at Lae and

Newton by beheading on 29 March at Salamaua. The latter's execution was by the Samurai code, using a sword. This has occasionally been touted as an 'honourable' death, as if it were some kind of privilege. It was not – it was murder, pure and simple, as was Lyon's bayoneting to death.

The squadron knew nothing of the fate of A28-3's crew until the following September when Salamaua was recaptured and Newton's body found. Lyon's body was not found until 1948.

Positive identification of Newton's body was a problem, but this was finally established in October, and on the 20th of that month he was posthumously awarded the Victoria Cross. The citation said in part: "Flt Lt Newton's many examples of conspicuous bravery have rarely been equalled and will serve as a shining inspiration to all who follow him". This could have been applied equally to many members of the squadron.

Flt Lt Bill Newton and his crew, Flt Sgt Lyon and Sgt Eastwood. (via Neil Mackenzie).

A Change of Scenery

In July 1943 No 22 Squadron moved from Port Moresby to Vivigani Strip on Goodenough Island in the Solomon Sea where it was joined by 30 Squadron's Beaufighters, 77 Squadron's Kittyhawks and 79 Squadron's Spitfires. The Boston and Beaufighter squadrons formed part of the RAAF's No 77 (Attack) Wing, itself a component of what would later be called the 1st Tactical Air Force (1 TAF), formerly No 10 (Operational) Group. This and other RAAF units in the area became part of the force which would fight in the campaign to recapture New Britain.

The township of Rabaul – the central string point of Japanese power in the area and a major base – was the key to success, and many air raids were conducted by the Allies. Without Rabaul, Japanese supply lines would be severely compromised. The Allied invasion of New Britain began in December 1943, and by the following March the Japanese base at Rabaul was a mere shadow of its former self. In addition, Japanese fighter opposition in the area had noticeably reduced by then.

From Goodenough, 22 Squadron's operations were similar to before, except now the flights to and from the targets

Homeward bound off the south-west coast of New Guinea, with Bob Wines' A28-6 'Japanese Ghost Producer' in the foreground. (via Neil Mackenzie)

(mainly in New Britain) involved longish flights over water rather than the rugged terrain of New Guinea. Over the next few months the squadron would also lend support to Allied seaborne assaults on Lae, Salamaua and Finschhafen.

The squadron's first mission from Goodenough in late July was also the largest RAAF strike yet undertaken in the area and involved 63 aircraft attacking Gasmata on New Britain. The weather was bad but 22 Squadron's assigned task of attacking aircraft on the ground at Gasmata Strip and the wireless station was successfully carried out.

The two attack squadrons of 77 Wing built up a considerable record of success conducting sweep missions against the large numbers of barges radiating from Rabaul to Japan's other bases in the South-West Pacific area. Barge traffic was very heavy in the middle of 1943 but by the end of the year it had just about disappeared thanks largely to the efforts of Nos 22 and 30 Squadrons. Each squadron maintained a high 'kill' rate on barges – up to 100 each per month – during this period. These were supported by attacks on other targets with equal success.

But these successes were not without cost and four more Bostons of the original batch were lost in the second half of 1943, three of them on the same day, September 12.

On that day, four of 22 Squadron's Bostons conducted a barge sweep and bomb attack on Gasmata Island. A28-15 (Flt Lt H B Dawkins) crash landed in the

Boston A28-8 immediately after its crash landing on Vivigani Strip, Goodenough Island, in September 1943. Flg Off Harry Rowell and his crew escaped unharmed. The aircraft had had its hydraulics shot away, causing a difficult return to Vivigani. (RAAF Historical/ 22 Squadron Historical)

An A-20C Boston over the water during operations in support of the Rabaul offensive in 1943-44. From the squadron code, the aircraft is A28-30.

sea after being hit in one engine and despite all three crew members being seen in their dinghy immediately afterwards, were not heard from again. Boston A28-16 (Flt Sgt R G Riley) failed to return from the mission even though it was seen to leave the target area and A28-8 (Flg Off Harry Rowell) was hit in the cockpit and hydraulic system but managed to limp back to base where it made a crash landing.

This Boston took no further part in the war, but in 1987 its remains were recovered from Goodenough Island and shipped back to Australia where it is now undergoing restoration for the RAAF Museum.

This left the Royal Australian Air Force with just six intact Bostons on its books, although replacements would begin arriving within days in the shape of nine A-20Cs. In November, a similar number of A-20As were taken on charge by the RAAF.

The replacement A-20Cs (A28-23 to 31) required substantial work to make them fully serviceable, resulting in them not conducting their first missions until mid October.

One of the new aircraft was lost shortly after arriving. On 12 October, A28-26 crashed and burnt on takeoff from Goodenough Island on its way out to bombing practice. The incident is vividly remembered by George Brain, a member of the salvage gang from 26 RSU (Repair and Salvage Unit). At the time of the crash Mr Brain was on duty alongside the control tower and recalled the scene in a letter to *Australian Aviation* magazine:

"At approximately the tower, the nosewheel strut on '26' folded up. The aircraft slid along the strip and overturned in a shallow ditch at the end of the strip and then caught on fire.

"An RAAF guard was on duty to halt traffic (trucks etc) crossing the end of the strip during ops. I cannot recall if he was from our unit – he was practically alongside 26! When the first bomb went off we were approximately 50 metres away.

A28-8 after recovery and stripping of useful components. The Boston's remains are now in Australia and undergoing restoration by the RAAF Museum. (D Hoskings via 22 Squadron Historical)

A-20A Boston A28-34 photographed from one of its 22 Squadron brothers. This aircraft was converted to components in February 1945. (RAAF)

"As the escape hatch was above the cabin and sitting on the ground, nothing could be done for the crew. There were five distinct explosions of the eight 100lb bombs on board. Out in the ocean were some rocks used by the Bostons and Beaufighters for bombing and strafing practice, hence the 100 pounders ..."

More losses were recorded between then and late November when the squadron moved to its new base on Kiriwina Island to the north and further into the Solomon Sea. One of the losses was A28-29 (one of the newly delivered aircraft) flown by the squadron's commanding officer, Wng Cdr W E Townsend. His aircraft was hit by anti-aircraft fire during a raid on Wuning Plantation and ditched in Jacquinot Bay, New Britain. The aircraft was seen to sink and the crew was assumed missing until they turned up a month later after an amazing adventure which included paddling to shore and then being passed from village to village by friendly natives, often right under the noses of the Japanese.

Wng Cdr J Emerton assumed command of the squadron when Townsend went missing.

An Eventful Year

The end of November and the move to Kiriwina marked the end of 22 Squadron's first year of active service in the New Guinea area. The statistics were impressive: 547 sorties carried out; 725,230lb (328 tonnes) of bombs dropped; 606,550 rounds of ammunition expended. Against this was the loss of 12 aircraft on operations, 13 personnel killed, 10 missing and six wounded.

Perhaps the following article from *Wings* (February 1944) best sums it up. Entitled "Boston Boys", the piece starts with Official War Correspondent, Ken-

Boston A28-21 'The Leprechaun', the victim of a takeoff crash and resultant fire at Vivigani in October 1943. (22 Squadron Historical)

neth Slessor, talking about a particular Boston:

"There was once a Boston bomber called *She's Apples* [A28-9] and she had a very good story ...

"For this bomber, famous throughout the SW Pacific, is claimed the bombing record for any single aircraft in this area. They painted the name on her old nose one day in honour of Wng Cdr C C Learmonth, DFC, and his WAG, Buzz Burn. Whenever they returned from an operation in her and someone asked Burn how the Boston had behaved, his invariable answer was; 'She's apples'.

"In a year crowded with the adventures of 186 sorties, she has been hit by the enemy only twice. Her life has been just as charmed as that of Wng Cdr Learmonth, who flew most of his operational trips in her. *She's Apples* was Learmonth's pet. He started as a Flight Lieutenant in her and finished with her as a Wing Commander at the age of 23. She carried him unharmed through storms of low level AA fire and brought him back from desperate encounters. She was never hit while Learmonth flew her". [Learmonth was killed during January 1944 in Western Australia].

The article goes on to summarise 22 Squadron's first year of operations in New Guinea:

"The Australian Boston squadron has completed its first year. A young squadron, but becoming legendary. It started with the handicap of replacement difficulties in equipment and personnel and it lost three machines and crews right at the beginning. Its aircraft flew in all weathers, almost ramming bullets down the enemy AA barrels. And the ground staff achieved miracles in improvisation and repair. Bostons staggered home almost shot to pieces but the maintenance staffs had them flying again.

"Remember that Bostons don't carry navigators – pilots do their own navigation. Yet those Bostons reached targets which other aircraft's navigators could not reach because of the weather. On the first 135 missions the squadron never once failed to reach and attack its objective ...

"Flt Lt W E Newton won the VC, the first RAAF VC in the Pacific Area. Wng Cdr K M Hampshire won a DSO, while Sqdn Ldr Learmonth, Flt Lt R A Wines, Flt Lt V H Hunt and Flt Lt P C Mullins won DFCs. Flt Sgt D O Duncan, a WAG, was mentioned in despatches. Further tribute was the selection of Flt Lt C R Sladen and Flt Sgt A H McKay to guide a formation of 50 Liberators and Mitchells to a target in the Mubo-Komiatum area.

"The Bostons first made news in Army co-op, but lost three in the first fortnight [the fragmentation bomb affair]. They spent two months tree-hopping over Buna and Salamaua bombing and strafing an invisible enemy – they bombed from photographs. Later the AIF found 250 dead Japs in one village.

"Boston boys are high in the military estimation. Isolated AIF troops thanked

'She's Apples', Boston A28-9/DU-K usually flown by Charles Learmonth and with 75 mission symbols painted on the nose. This Boston was previously known as 'Pegasus'. (via Neil Mackenzie)

Sweeping in low with bomb doors open, 22 Squadron Bostons prepare to attack a Japanese position during 1943. (via James Grant)

them for tobacco, food and supplies [which had been dropped to troops during the Mubo assault]. Grateful troops sent the squadron a Jap light machine gun and a mortar as souvenirs. Prominent in these operations were Flg Off J C Miles and Flt Sgt F Dunn; Flg Off H M Rowell and Flt Sgt D Lyons; Flg Off L Kenway and Flt Sgt F Gordon; Flg Off J McMaster and Flt Sgt J Larkin; Flt Lt C R Sladen and Flt Sgt A G Smylie; Plt Off G T Smith and Sgt R Kerr; and Flt Sgt E Riley and Flt Sgt L Wilson.

"A Boston formation located five Jap destroyers embarking reinforcements for Buna. A destroyer was promptly crippled (a Fortress sank it) before the Bostons started on the landing barges, strafed the swimming troops and blew up two big dumps.

"The bad weather after the fall of Buna did not stop the Boston Boys strafing Jap installations and airstrips. One Boston over the Owen Stanleys spun out of control in a downdraught from 14,000ft to 500ft. The pilot force landed in a clearing near Wairope Bridge. Flt Lt Sladen, Flg

Wng Cdr J Emerton (left) with Frank Bagnel (war photographer) Flg Off T Gawne in front of A28-9 'She's Apples'. Emerton was 22 Squadron's commanding officer in November 1943 was missing in action just two months later. (22 Squadron Historical)

No 22 Squadron moved to Morotai Island in the late 1944, allowing strikes to be carried out even as far north as the Philippines. On Morotai, A-20C Boston A28-28 noses up to one of the original batch, possibly A28-9 'She's Apples'. (22 Squadron Historical)

Deliveries of the improved A-20G Havoc/Boston began in June 1944, providing 22 Squadron with enhanced capability. Here, Flg Off R Wemyiss and Plt Off R S Woodgate examine one of the 'G' model's improved features, its power operated upper turret. (via Neil Mackenzie)

Off Chester and Sgts Williams and Kendall tramped two days from Kokoda, stripped the plane and cut a runway out of the jungle. Sladen just skimmed the leaves in his takeoff. The others walked.

"Bostons were in the Bismark Sea battle, starting fires on Lae to ground the fighters, hitting a destroyer and generally helping. In this were Sqdn Ldr Learmonth and Flt Sgt K Crowe; Flg Off Skinner and Flt Sgt K Pritchard; Flt Lt H B Craig and Flt Sgt J Calthorpe; Flt Lt Wines and Flt Sgt C Grove; Flt Lt H V Hunt and Flt Sgt J Grant.

"Soon after, the Bostons made smashing low level attacks on Salamaua airfield and dump areas. A 40,000 gallon petrol tank went up. It was in this phase that Newton won his VC ...

"Flt Sgt Duncan, mentioned in despatches, typifies the Boston Boys. In an attack on a wrecked cargo ship the Japs were using as a supply depot, his foot was almost shot off and a gunner killed but he stuck to his wireless. In the stretcher after he landed he grinned, 'I guess this will be a homer'".

More Moves and New Equipment

No 22 Squadron spent the first half of 1944 operating from Kiriwina, continuing with its sorties against Japanese barges, troop positions, stores depots, bridges and roads. The successes continued but so did the losses, including that of the CO, Wng Cdr Emerton, shot down in A28-27 whilst photographing gun positions at Kabu River in late January. Sqdn Ldr (shortly Wng Cdr) J Hickey assumed command. Hickey remained CO until August when Sqdn Ldr C Woodman took over. He was the final commanding officer of No 22 Squadron when it was equipped with Bostons, and his command extended into the Beaufighter era.

By the middle of the year, the Allies were steadily pushing the Japanese back and reclaiming islands as they went. Noemfoor Island off Geelvink Bay on the north-eastern coast of Dutch New Guinea was taken in early July and immediately became a substantial RAAF base, including for No 22 Squadron which was established there by early August.

Meanwhile, in June, the squadron had received the first of 29 A-20G Havocs/Bostons which would appear on RAAF books (A28-50 to 78). This more capable version of the Boston with its power operated mid-upper turret and ability to carry ordnance under its wings as well as in the fuselage bomb bay represented a useful upgrading in capability.

The squadron's main tasks from Noemfoor were to harass enemy shipping and attack airfields in the area, specifically those at Otiwiri, Sagan and Mongasan. These were attacked over three days at the end of August in 47 sorties during which 46,500lb (21 tonnes) of bombs were dropped and the airfields put out of action.

Other operations during August 1944 included sorties against shipping and ground targets including the Japanese headquarters on Soanek Island. Despite only starting operations from Noemfoor in the middle of August, the squadron recorded 26 operations and 106 sorties during that month, without loss.

September was a different story, with four aircraft lost, three on operations and one during a training flight. The squadron's efforts during that month were concentrated on supporting the Allied push to Morotai Island in the Halmaheras group, a vital link in the chain back to the Philippines. Barge sweeps and attacks on general shipping occupied the squadron for much of September, as it did in October. By then, Morotai was in the Allies' hands and was being established as a major base with its twin airstrips, Pitu and Wama.

In mid November 1944, the constituent units of No 1 Tactical Air Force – including 22 Squadron's Bostons – began moving into Morotai. 22 Squadron was established at Wama Strip by the 17th and on the following day carried out strikes against enemy airfields, watercraft and installations in the Northern Celebes, all in support of the American landings in the Philippines.

Armourers hard at work replenishing the nose guns of a 22 Squadron Boston with tracer ammunition. Working conditions were always difficult, particularly in New Guinea. (22 Squadron Historical)

22 Squadron personnel pose for the photographer on Morotai Island. The aircraft is an unidentified A-20G. (via Neil Mackenzie)

Bombing up an A-20G while some of the squadron's aircrew, at left, discuss plans for the forthcoming raid. They are (from left) Wnt Off L C Fanning, Sgt R Cameron, Flt Lt C E Spence and Flg Off G T Dick. The photograph was taken at Noemfoor in October 1944. (via Neil Mackenzie)

A28-68 (or possibly -66; both carried a 'P' code), an A-20G Havoc, undergoes some repair work at Morotai, including some fuselage reskinning. (via Neil Mackenzie)

The end of RAAF Boston operations, the result of a Japanese raid on Morotai on the night of 22 November 1944. The raid damaged 13 Bostons, all but two of them beyond repair. This pile of junk represents four aircraft! No 22 Squadron re-equipped with Beaufighters for the duration of the war. (RAAF Historical)

Boston Finale

No 22 Squadron's Bostons made a small piece of history on 22 November when they became the first RAAF aircraft to strike as far north as the Philippine Islands, hitting Bunawan Harbour on Mindanao Island. But any celebrations were short lived, as on the same night, one of the numerous 'nuisance' raids on Morotai by the Japanese effectively ended the Boston's career in the RAAF.

The raid damaged 13 aircraft – A28-6, 10, 51, 52, 54, 57, 59, 60, 65, 67, 68, 75 and 76 – of which only two (54 and 57) were not judged to be beyond repair and converted to components.

The raid decimated the squadron's strength to the extent that it was immediately withdrawn to Noemfoor for re-equipment with Beaufighters. As 22 Squadron was the only RAAF unit operating Bostons, it was a bit of an 'orphan' in one sense, and since no replacement Bostons were readily available and the unit's 'partner in crime' – 30 Squadron – was equipped with Beaufighters, the change was a sensible and practical one. Not that this necessarily pleased 22 Squadron's aircrew; they had a very high regard for the Boston, which offered performance, reliability and durability.

The surviving Bostons carried out miscellaneous duties including communications flights. On 28 December 1944, one of them recorded another piece of history by becoming the first RAAF aircraft to land at Leyte and Mindoro in the Philippines since their recapture by the Americans. The Boston was carrying mail for the troops.

In February 1945 No 22 Squadron returned to Morotai equipped with Beaufighters, and it was with these aircraft that it ended its active service in World War II.

The Boston did not remain in RAAF service for very long, the surviving A-20Gs being returned to the USAAF in early 1945 and the survivors of the three other marks were converted to components in the first few months of the year.

A perhaps inglorious end to a notable chapter of RAAF history.

RAAF DOUGLAS A-20 BOSTONS

Abbreviations: soc – struck off charge; USAAF – United States Army Air Force; USN – United States Navy; n/a – not available/applicable; w/o – written off; cvtd – converted; compnts – components; u/c – undercarriage; t/o – takeoff; u/s – unserviceable; a/c – aircraft; AA – anti aircraft.

Notes: Of the 22 DB-7B Boston IIIs (A28-1/22), 11 are Boeing built Boston IIIAs. These are A28-2, 10-13 and 15-20.

RAAF No	Model	Prev No	Deliv	Disposal/Remarks
A28-1	DB-7B	AL890	03/42	crashed on t/o Kiriwina 12/42
A28-2	DB-7B	AL347	03/42	crashed on t/o Richmond NSW 05/42
A28-3	DB-7B	AL887	03/42	ditched near Salamaua NG 03/43; a/c of Flt Lt W Newton VC
A28-4	DB-7B	AL893	03/42	damaged by enemy AA, crash landed Port Moresby 01/43, cvtd to components
A28-5	DB-7B	AL895	03/42	damaged in action Ring Ring plantation New Britain 08/43, cvtd to components
A28-6	DB-7B	AL897	03/42	damaged in air raid Morotai 11/44, cvtd to components
A28-7	DB-7B	AL899	03/42	crashed on t/o Gurney Strip Milne Bay NG 06/44
A28-8	DB-7B	AL907	03/42	remains extant RAAF Museum Point Cook Vic for restoration
A28-9	DB-7B	AL891	04/42	converted to components 02/45
A28-10	DB-7B	AL358	04/42	damaged in air raid Morotai 11/44, cvtd to components
A28-11	DB-7B	AL364	04/42	converted to components 02/45
A28-12	DB-7B	AL365	04/42	exploded and crashed Port Moresby Harbour 11/42
A28-13	DB-7B	AL367	04/42	crashed near Point Hood NG 06/43
A28-14	DB-7B	AL892	04/42	missing after raid on Malahange NG 02/43
A28-15	DB-7B	AL361	04/42	lost on ops Gasmata New Britain 09/43
A28-16	DB-7B	AL362	04/42	lost on ops Gasmata New Britain 09/43
A28-17	DB-7B	AL363	04/42	damaged by engine fire 05/42, cvtd to components
A28-18	DB-7B	AL366	04/42	crash landing Noemfoor 11/44, cvtd to components
A28-19	DB-7B	AL368	04/42	crashed on ferry flight near Woodford Qld 10/42
A28-20	DB-7B	AL369	04/42	blew up over target Buna NG 11/42
A28-21	DB-7B	AL894	04/42	shot down near Salamaua 02/43
A28-22	DB-7B	AL898	04/42	blew up over target Buna NG 11/42
A28-23	A-20C	42-33154	09/43	crash landed Kamiri strip Noemfoor 10/44, cvtd to components
A28-24	A-20C	42-33163	09/43	crashed in sea off Kiriwina 03/44
A28-25	A-20C	42-33172	09/43	cvtd to components 04/45
A28-26	A-20C	42-33180	09/43	crashed and burnt on t/o Goodenough Is 10/43
A28-27	A-20C	42-33134	09/43	lost in action Kubu River New Britain 01/44
A28-28	A-20C	42-33142	09/43	converted to components 05/45
A28-29	A-20C	42-33211	10/43	shot down into sea New Britain 11/43
A28-30	A-20C	42-33174	10/43	damaged by accidental discharge of dorsal gun Kamiri 08/44, cvtd to components
A28-31	A-20C	42-33168	11/43	converted to components 05/45
A28-32	A-20A	40-85	11/43	converted to components 04/45
A28-33	A-20A	40-143	11/43	converted to components 05/44
A28-34	A-20A	40-3160	11/43	converted to components 04/45
A28-35	A-20A	40-162	11/43	converted to components 03/45
A28-36	A-20A	40-77	11/43	converted to components 04/45
A28-37	A-20A	40-118	11/43	converted to components 04/45
A28-38	A-20A	40-139	11/43	converted to components 04/45
A28-39	A-20A	40-144	11/43	to instructional airframe, soc 10/46
A28-40	A-20A	40-3159	11/43	converted to components 04/45
A28-41 to A28-49 not allocated				
A28-50	A-20G	43-21296	06/44	crashed in sea near Tabo Is 09/44
A28-51	A-20G	43-21306	06/44	damaged in air raid Morotai 11/44, cvtd to components
A28-52	A-20G	43-21308	06/44	damaged in air raid Morotai 11/44, cvtd to components
A28-53	A-20G	43-21311	06/44	missing on ops 09/44
A28-54	A-20G	43-21540	06/44	damaged in air raid Morotai 11/44
A28-55	A-20G	43-21303	06/44	lost in action Kai Island 09/44
A28-56	A-20G	43-21314	06/44	issued to USAAF 03/45
A28-57	A-20G	43-21301	06/44	damaged in air raid Morotai 11/44, issued to USAAF Biak 01/45
A28-58	A-20G	43-21294	06/44	issued to USAAF 01/45
A28-59	A-20G	43-21542	06/44	damaged in air raid Morotai 11/44, cvtd to components
A28-60	A-20G	43-21545	06/44	damaged in air raid Morotai 11/44, cvtd to components
A28-61	A-20G	43-21380	07/44	missing on test flight from Noemfoor 09/44
A28-62	A-20G	43-21381	07/44	brake failure on landing Leyte 01/45, hit parked USN Corsairs, issued to USAAF
A28-63	A-20G	43-21384	07/44	crash landing Kamiri Strip Noemfoor 09/44, cvtd to components
A28-64	A-20G	43-21387	07/44	issued to USAAF 01/45
A28-65	A-20G	43-21953	08/44	damaged in air raid Morotai 11/44, converted to components
A28-66	A-20G	43-21976	08/44	issued to USAAF 01/45
A28-67	A-20G	43-21977	08/44	damaged in air raid Morotai 11/44, cvtd to components
A28-68	A-20G	43-21985	08/44	damaged in air raid Morotai 11/44, cvtd to components
A28-69	A-20G	42-54069	09/44	issued to USAAF 01/45
A28-70	A-20G	42-54091	09/44	issued to USAAF 01/45
A28-71	A-20G	42-54073	09/44	issued to USAAF 01/45
A28-72	A-20G	42-54092	09/44	issued to USAAF 01/45
A28-73	A-20G	42-54098	09/44	issued to USAAF 01/45
A28-74	A-20G	42-54120	09/44	issued to USAAF 01/45
A28-75	A-20G	43-21970	10/44	damaged in air raid Morotai 11/44, cvtd to components
A28-76	A-20G	43-21978	10/44	damaged in air raid Morotai 11/44, cvtd to components
A28-77	A-20G	43-22250	10/44	issued to USAAF 02/45
A28-78	A-20G	43-22140	10/44	issued to USAAF 02/45

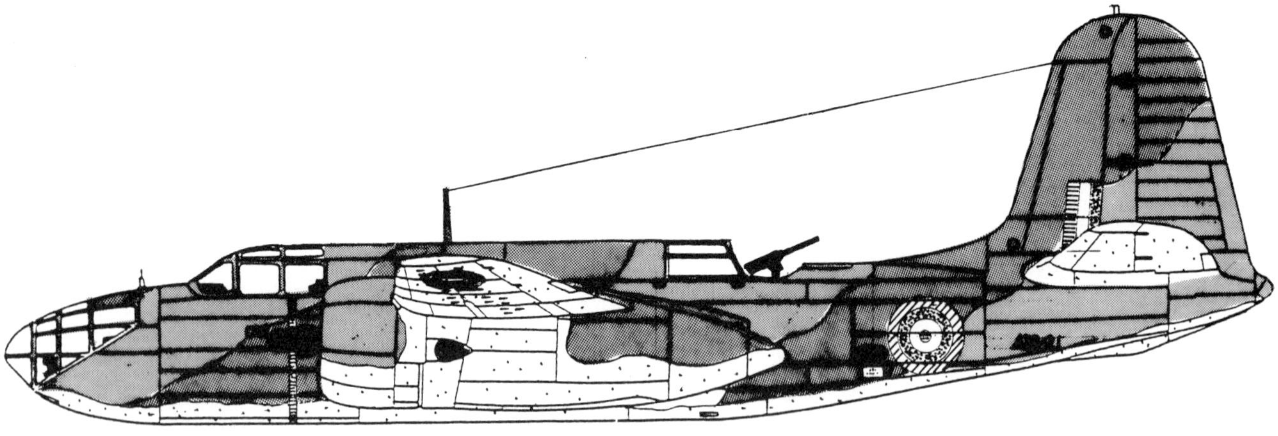

Boston III A28-21 shortly after delivery
Dark green and dark earth upper surface with sky type 'S' undersurface. Black serial. Red propeller warning stripe below cockpit. Black propeller hub.

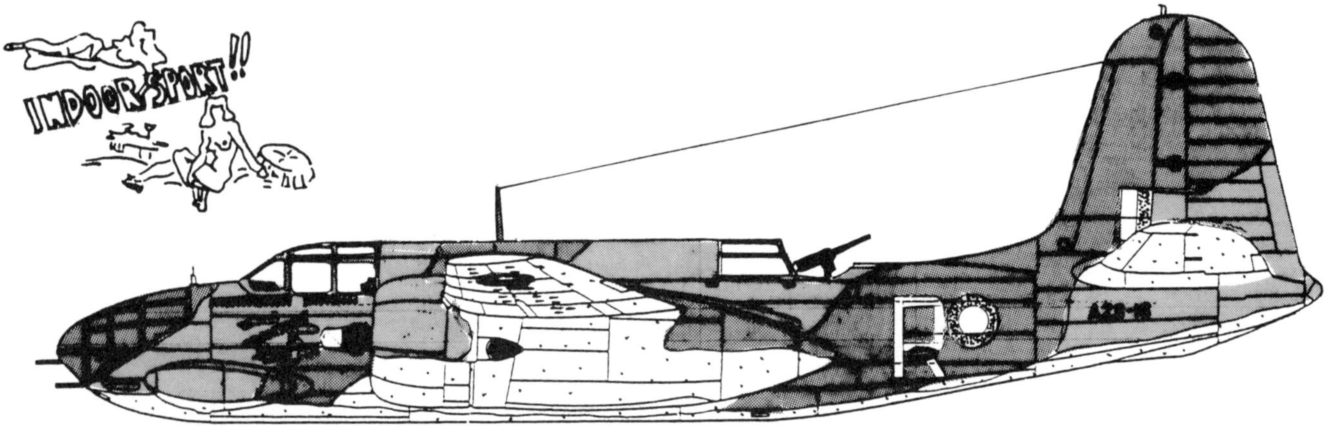

Boston IIIA A28-16/R of No 22 Sqn, Port Moresby, November 1942
Dark green and dark earth upper surface with sky type 'S' undersurface. Light grey code. Black serial. Light grey name and artwork. Light grey bombing mission marks below cockpit divided into two groups (nineteen in left group and eleven in right group). Natural metal propeller hub.

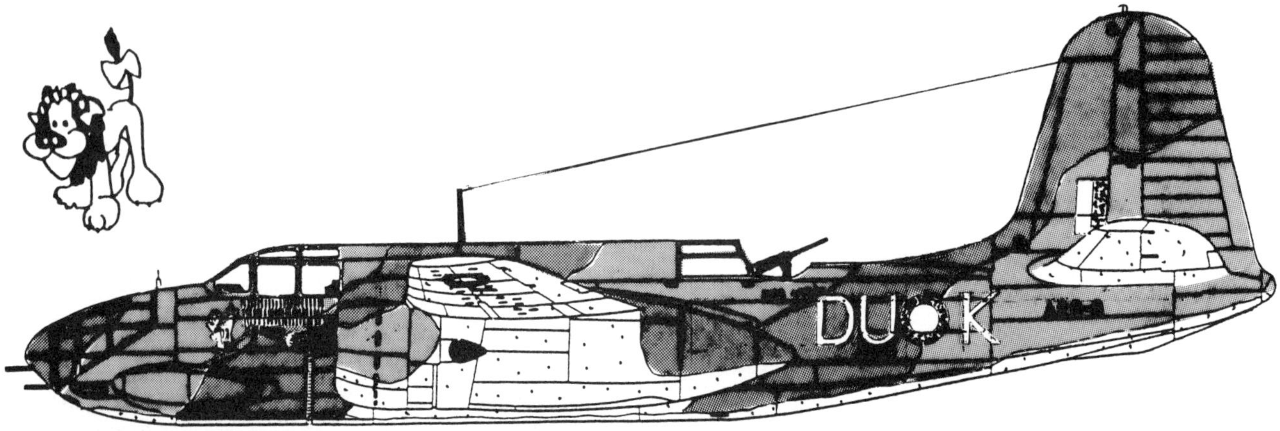

Boston III A28-9/DU-K of No 22 Sqn, Goodenough Is, 1943
Dark green and dark earth upper surface with sky type 'S' undersurface. Light grey code. Black serial. Red propeller warning stripe below cockpit. Natural metal showing through on black propeller hub. Black "SHE'S APPLES" below rear gun emplacement. Light grey bombing mission marks with thin black outline in four rows from top – 24, 23, 22, 6). Light brown lion caricature on nose, with blue ribbons on head and tail, black mane, tail, details and outline.

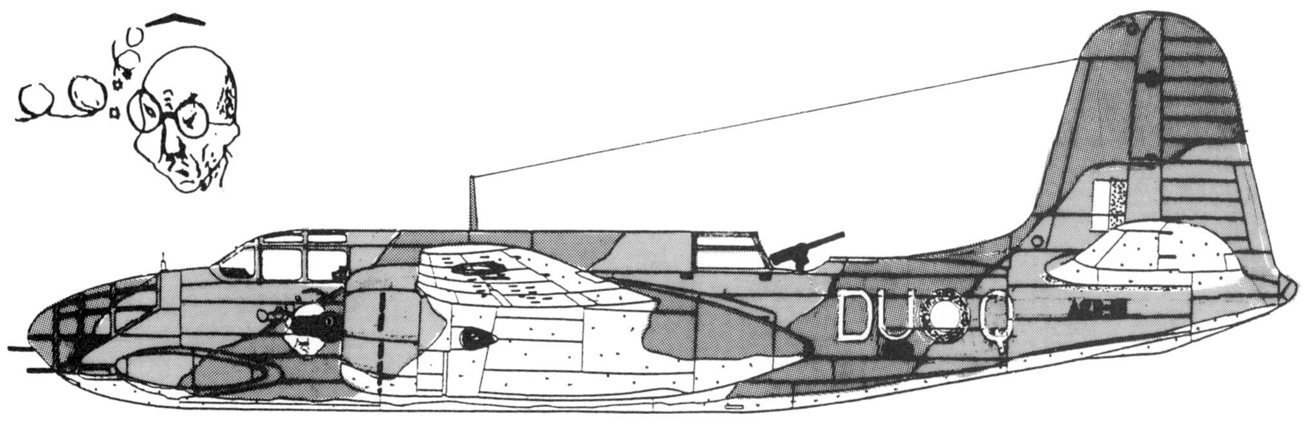

Boston III A28-15/DU-Q of No 22 Sqn, Goodenough Is, 1943
Dark green and dark earth upper surface with sky type 'S' undersurface. Light grey code and serial. Yellow face below cockpit with black details, brown boomerang and black swirls.

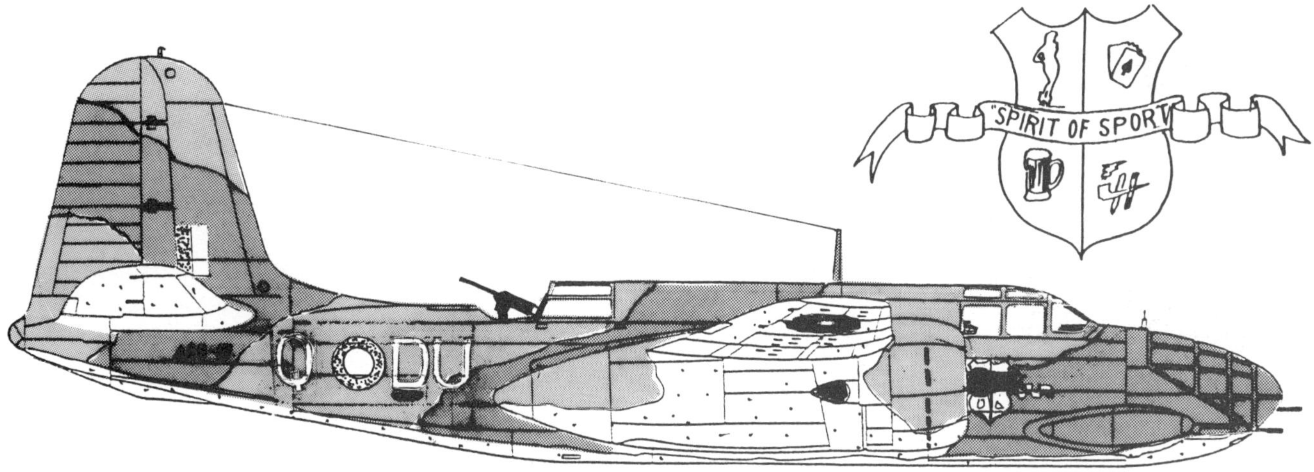

Right side of **A28-15/DU-Q** *showing the standard camouflage pattern on this side. Artwork below cockpit consists of a dark blue shield and banner with white outline, lettering, figure, cards and beer mug, brown pipe, cigar and filter tip and white cigarette.*

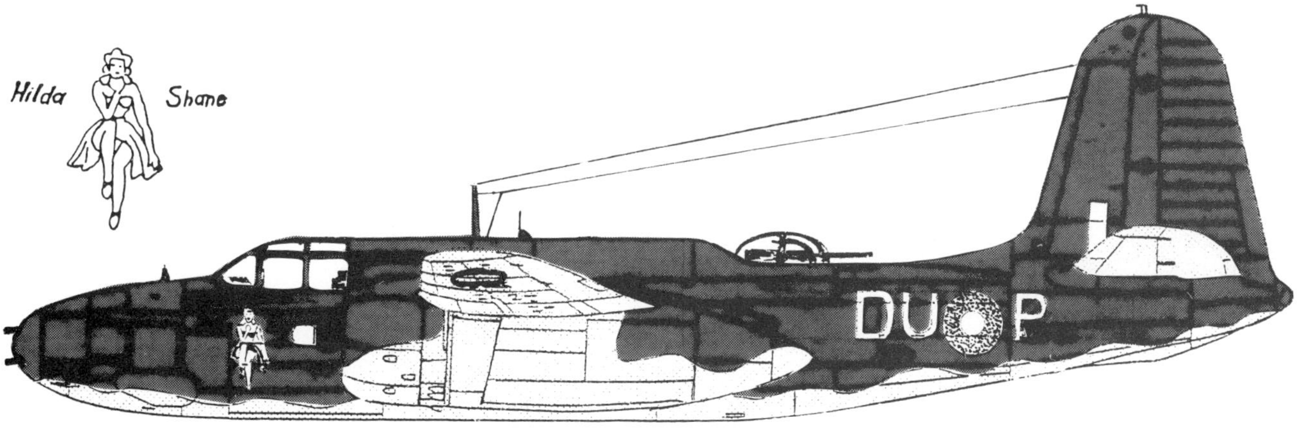

A-20G A28-60/DU-P of No 22 Sqn, Noemfoor, 1944
Olive drab upper surface with neutral grey undersurface. White code and serial. Only white half of fin flash appears on the fin. Natural metal propeller hub. Figure below cockpit is flesh with red lips, yellow hair, blue dress with white shading and thin black outline and details. White "Hilda Shane". Seventeen bombing mission marks in white.

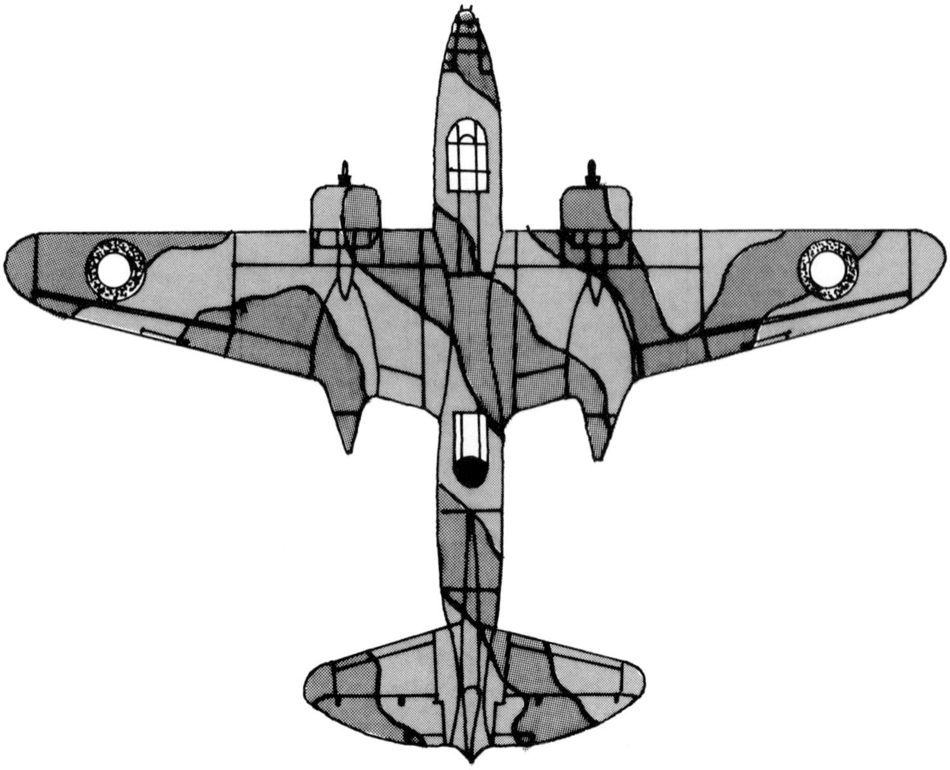

Standard upper surface camouflage pattern

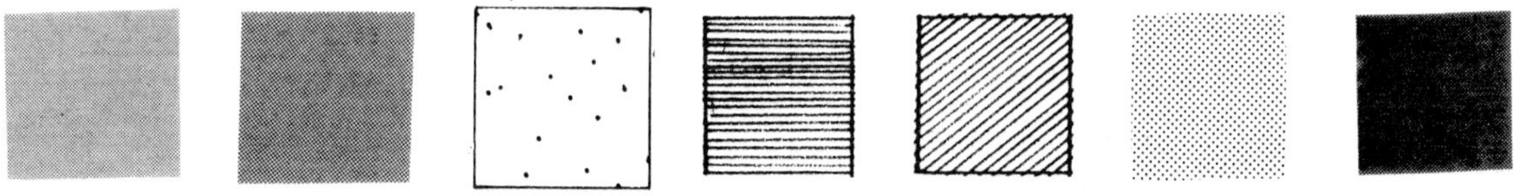

Dark Earth	*Dark Green*	*Sky Type 'S'*	*Red*	*Yellow*	*Neutral Grey*	*Olive Drab*

NORTH AMERICAN
B-25 MITCHELL

Interesting angle on the Australian War Memorial's TB-25J Mitchell flying in Australia before it was put into long term storage, a great shame for an airworthy warbird. The Mitchell was purchased from a private consortium in 1983 after it had been brought into Australia from the USA. It flew until 1987 before being stored. Originally in US Navy colours, the aircraft acquired quasi-RAAF marks as 'A47-31' of No 2 Squadron. The real A47-31 was a B-25J which was scrapped in 1950. The TB-25J was originally flown with the US serial number 44-86791.

NORTH AMERICAN B-25 MITCHELL

Before 1940, North American Aviation of Inglewood, California, was not an aircraft manufacturer which could be considered a major producer of fine combat aircraft. Between its formation in 1928 and World War II, the company's products of note included the O-47 single engined observation aircraft and the NA-16 series of trainers which developed into the T-6 Texan/Harvard family, built in very large numbers and used as the Allies' standard advanced trainer during the war. By the end of the conflict, North American Aviation had designed and built in significant quantities two of the great combat aircraft of the age, the P-51 Mustang fighter and the B-25 Mitchell medium bomber.

The company's strength was in its management: president and general manager J H 'Dutch' Kindelberger, vice president Leland Atwood and vice president in charge of engineering Raymond Rice. This formidable trio was able to transform North American into the significant aircraft manufacturer it became through the Texan, Mustang and Mitchell and equally importantly, was able to sustain that momentum after the war with another classic combat aircraft, the F-86 Sabre jet fighter.

In the case of both the Mustang and the Mitchell, North American had little or no previous experience with such advanced combat designs, and in both aircraft the result was arguably the best of their type in World War II. Design of what would become the Mitchell preempted that of the Mustang by two years, so the medium bomber was the company's first experience of new aspects of aircraft design: a bomber, a twin engined type and a high performance warplane of any sort. There had been a couple of paper projects, but no hardware.

For the Royal Australian Air Force, the Mitchell was numerically a relatively unimportant type with just 50 being taken on strength from 1944 and serving with only one operational squadron. But Australia's involvement with the aircraft was greater than that through the Mitchells operated in Australia by No 18 Squadron Netherlands East Indies Air Force. Comprising a mixture of Dutch and Australian personnel, the squadron operated within the framework of the RAAF (as did other NEI squadrons) and formed the RAAF's No 79 Wing along with No 2 Squadron RAAF, which swapped its Beauforts for Mitchells from 1944.

The NA-40

The B-25 Mitchell's evolution began in early 1938 with the issuing of a United States Army Air Corps proposal for a twin engined medium bomber. North American came up with the abortive XB-21 but followed this with the private venture NA-40, a late starter in the attack bomber competition which resulted in several types making it into production, the most notable being the Douglas Model 7B which was developed into the A-20 Havoc/Boston family. It was under the terms of this competition that the prototype NA-40 was built, along with other types from several manufacturers.

The NA-40 was a rather angular looking three seat twin engined shoulder wing monoplane powered by two 1,100hp Pratt & Whitney R-1830 Twin Wasp 14 cylinder two row radial engines. It featured twin fins and rudders and its appearance was dominated by the large 'greenhouse' canopy under which the pilots sat in tandem. Ahead of that was a bulbous transparent nose for the navigator/bombardier. Design maximum bomb load was 1,200lb (544kg) as per the Army Air Corps' specification and defensive armament comprised a single 0.30in machine gun on a flexible mount in the nose, although the design provided for wing, ventral and dorsal gun positions.

The NA-40 was of modern all metal construction and like the Douglas design for the same specification, featured a then radical departure from the norm, a tricycle undercarriage. Wing span was 66 feet (20.1m), gross weight 19,500lb (8,845kg) and maximum speed 265mph (426km/h).

The NA-40 (sometimes referred to as the NA-40-1) recorded its maiden flight in January 1939 with company test pilot Paul Balfour at the controls. First impressions were that the NA-40 was seriously underpowered and steps were quickly taken to rectify that situation.

The aircraft was re-engined with 1,350hp Wright GR-2600 Double Cyclone 14 cylinder radials almost immediately and reflew with these powerplants installed in mid February 1939 as the NA-40-2. It was immediately 20mph

A Mitchell on short finals displays some of its characteristics – the 'gull' wing, the tricycle undercarriage and the twin fin tailplane design. (via Neil Mackenzie)

An early production B-25 Mitchell in United States Army Air Corps colours. Deliveries began in the second half of 1940 and by the time production ended in August 1945, just under 10,000 of all versions had been manufactured.

(32km/h) faster, but more than that, the NA-40 represented pretty much the state-of-the-art of the American aircraft industry at the time. With the extra power came an increase in maximum weight to 21,000lb (9,525kg) and in March 1939 the transfer of the aircraft from NAA's facility in Los Angeles to Wright Field for evaluation by the Army Air Corps.

As had happened with the prototype Douglas Model 7B just two months earlier, the North American NA-40 was to have a very short life before being destroyed in a crash. Whilst turning on approach to land at Wright, Major Younger Pitts lost control of the aircraft and it crashed, fortunately without serious injury to anyone. The NA-40 was destroyed by fire, but not before its virtues and potential had been appreciated by the Army Air Corps. The crash was found to be not any fault of the aircraft or its design and its development was authorised to continue.

The B-25 Evolves

This authorisation to continue came with a substantial list of design improvements as suggested by the Air Corps along with a revised specification which took the NA-40 out of the 'attack' class

(right) Colonel William Mitchell, the staunch advocate of air power after whom the B-25 was named. Mitchell suffered for his unpopular advocacy during the 1920s and was even court-marshalled. Events proved his theories correct and he was posthumously promoted to the rank of Brigadier-General.

The Royal Air Force operated the Mitchell in large numbers, beginning with the similar B-25C (above) and B-25D (below), both designated Mitchell II in British service. Both shots provide some detail, the lower capturing the moment of exit of a stick of 500lb bombs. (via Neil Mackenzie)

Another RAF Mitchell II, this time with its Bendix ventral gun turret extended. (via Neil Mackenzie)

and into the 'medium bomber' category. North American therefore began an extensive redesign of its bomber under the leadership of Atwood and Rice and under the company designation NA-62. The result was the B-25 Mitchell as we now know it.

The redesign revolved around the new specification's requirement for the doubling of the original bomb load from 1,200lb (544kg) to 2,400lb (1,088kg) all carried internally in a lower fuselage bomb bay. The fuselage was widened, the original raised cockpit position was lowered and its 'roofline' made flush with the top fuselage line, the wing was attached to a point on the fuselage several feet lower than before (giving improved mainspar location) and the engine nacelles were extended aft of the wing trailing edge. The wing remained essentially as before but with an increase in span to 67ft 7in (20.60m) and retained its constant dihedral from root to tip, a feature which would shortly disappear from the design.

With the modification of the cockpit's height and contours came changes to the nose shape and transparencies design, the twin fins and rudders were given greater chord at their lower ends and the widened fuselage allowed pilot and copilot to sit side by side in the cockpit. Power was provided by two Wright R-2600-9 Double Cyclones each of 1,700hp and the design maximum weight increased to 27,310lb (12,388kg).

The new design featured a crew of five and defensive armament comprised three 0.30in machine guns, manually operated, in the nose and waist (one on each side), plus a 0.50in gun in a new extended tail cone. The bomb load was normally 2,000lb (907kg) but a maximum of 3,600lb (1,633kg) could be carried, well in excess of the Army's requirements.

Structurally, the NA-62 featured a cantilever wing comprising a two spar centre section which was permanently attached to the fuselage and contained integral fuel tanks and the engine mounts/nacelles, single spar outer wing sections and detachable wing tips. The fuselage was of semi monocoque aluminium alloy construction and the single wheeled tricycle undercarriage retracted into the lower nose section and engine nacelles.

The new design was approved by the Army Air Corps on 10 September 1939 (exactly one week after Britain and its Allies declared war on Germany) and ten days later North American Aviation received a $US11.7m contract for 184 aircraft, now officially dubbed the B-25.

The name 'Mitchell' would be officially applied in early 1941. The use of this name reflected a remarkable turnaround in American attitudes. Colonel William Mitchell had been court-marshalled in 1924 for his obsessive belief in the benefits of air power. His outspoken views had so annoyed his political 'superiors' that he was officially dealt with, but time proved him correct to the extent that he had an Army Air Corps bomber named after him and he was posthumously promoted to the rank of Brigadier-General. It was Mitchell who wrote in 1924: "Future wars will be fought in the sky and only air forces can cope with the threat of enemy planes and submarines. Surface navies are useless because the most modern battleship can be sent to the bottom by aerial bombing". It's a pity Mitchell didn't live to see his views vindicated to a very large extent.

There was no XB-25 Mitchell prototype as such and the first of 24 B-25s scheduled to be delivered to the USAAC in 1940 took to the air for the first time on 19 August. This had been preceded by a static test airframe. Testing revealed a high performance bomber with a maximum speed of 322mph (518km/h) but also a lack of directional stability, particularly in the bombing run. The wing's constant dihedral was found to be the primary cause, resulting in all Mitchells from aircraft number ten onwards having their outer wings re-rigged for zero dihedral, giving the aircraft its distinctive 'gullwing' shape. In addition, the fin and rudder assemblies were given slightly more area by squaring them. The designation 'B-25' remained unchanged for the modified aircraft, and it covered the first 24 aircraft built.

From early 1941 production switched to the first of 40 B-25A Mitchells, with modifications incorporated which reflected information gleaned from the European air war. These included self sealing fuel tanks and additional armour protection for the crew, albeit at the cost of fuel capacity which reduced the aircraft's range by a substantial margin. The need for increased defensive armament was also highlighted by the goings on in Europe, but the B-25A was identical to the original model in this respect.

By mid 1941 the Mitchell had entered operational service with the newly renamed United States Army Air Force (USAAF), the first combat outfit equipped with the new bomber being the 17th Bombardment Group at McChord Field, Washington. Several months of operational trials revealed the B-25 as being a 'friendly' aircraft for both the air and ground crews and of considerable potential value as a combat aircraft. After the Japanese strike on Pearl Harbour in December 1941 and the USA's entry into the war, the 17th BG's three squadrons were transferred to the Pacific coast for anti submarine duties.

It was there that the Mitchell recorded its first 'kill' on Christmas Eve 1941 when an aircraft piloted by Brick Holstrom of the 17th BG's 95th Bomb Squadron stumbled upon a Japanese submarine, attacked, and sank it with a direct hit on the conning tower.

The B-25B and Tokyo

Despite the Mitchell's speed and potency as a medium bomber, the problem of a relative lack of defensive armament was not addressed until the B-25B of 1941. One hundred and twenty B-25Bs were built, completing the original order for 184 aircraft. The B-25B was the first variant to serve with Britain's Royal Air Force, 23 of them going into service as the Mitchell I and used only for familiarisation and training.

The B-25B had its beam and ventral 0.30in machine guns replaced by dorsal and ventral electrically operated turrets each mounting a pair of 0.50in guns. The ventral turret was retractable and used a

Two more views of RAF B-25C Mitchell IIs, showing upper surface and undercarriage detail. The B-25C/D series introduced greater fuel capacity and increased defensive and offensive armament. The RAF took delivery of 335 Mitchell IIs. (via Neil Mackenzie)

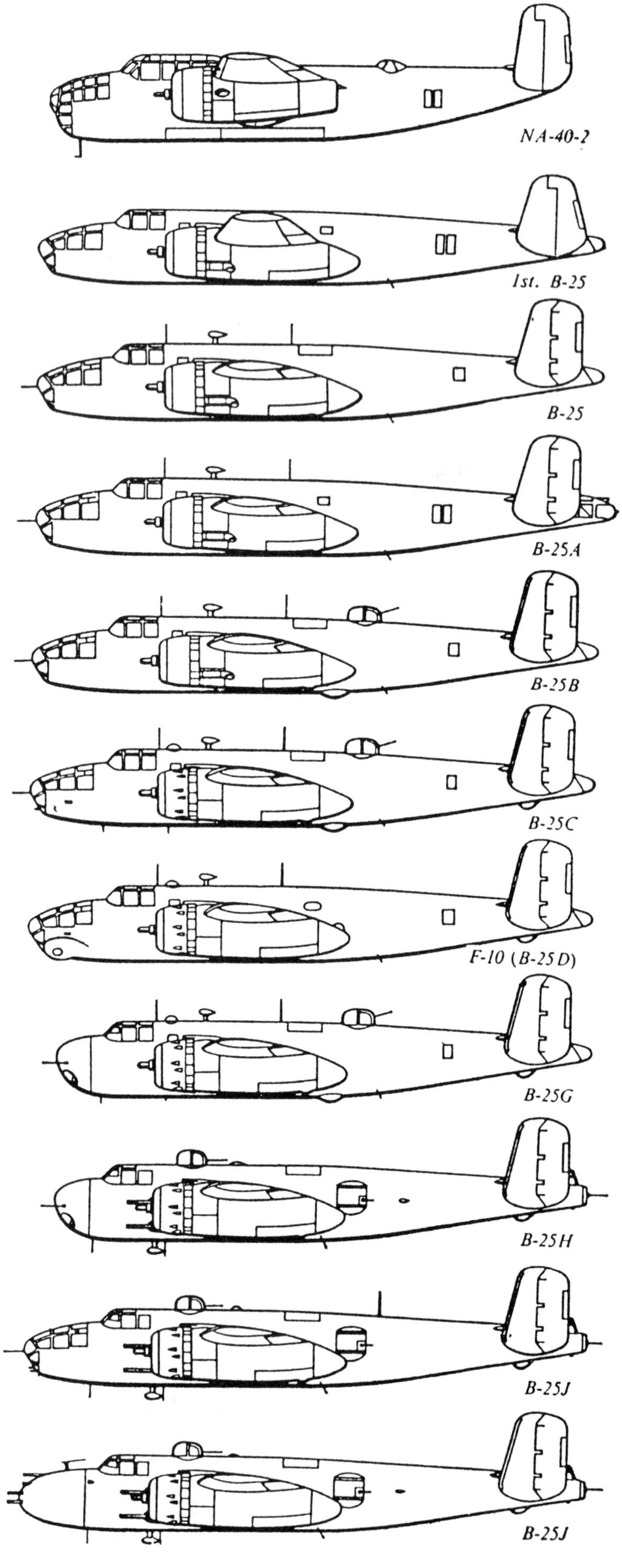

periscope for sighting by the gunner who kneeled above it. The nose gun remained as before but the tail gun was removed and its smaller housing was used as a prone observation post. A separate photographic station was located between the dorsal turret and the tail and maximum weight was increased to 28,640lb (12,991kg).

The Mitchell became famous through its participation in one of the best known and most audacious raids of the war. On 18 April 1942, sixteen B-25Bs were launched from the carrier USS *Hornet* and struck targets in Tokyo, Yokohama, Nagoya and Kobe. It was the first time the Japanese mainland had been attacked, and although the raid was of minimal military value, it did wonders for the morale of the American people and for others whose people had felt the onslaught of the Japanese.

Until then, the Japanese had seemed almost invincible as they achieved their ambitions throughout the Pacific.

The choice of aircraft for the raid lay between the Mitchell and the Martin B-26 Marauder, the former being selected for its superior airfield performance. Under the leadership of Lt Col James Doolittle (who among other achievements had won the 1925 Schneider Trophy race for the USA in a Curtiss CR-3), volunteer crews were selected from the 17th Bombardment Group and the 89th Reconnaissance Squadron and trained in secret at Hurlburt Field in Florida. Much of the training concentrated on taking off in the very short distance provided by the *Hornet's* deck and the successful achievement of this was a major feat in itself. As a result, the Mitchell became the heaviest aircraft to be launched from a carrier up to that time.

The B-25Bs were considerably modified for the raid. The ventral turret was removed (saving 600lb weight) and the space filled with extra fuel tanks which gave the aircraft a total capacity of 950imp gal (4,319 litres), compared with the standard 558imp gal (2,536 l). The raid was to be conducted from the low altitude of 1,500 feet so a special low level bomb sight replaced the still secret Norden sight. Two wooden dummy machine guns were installed in the tail to discourage attack from the rear by Japanese fighters and the takeoff weight for the mission went up to 31,000lb (14,062kg). Despite this, training revealed that by using special techniques, a Mitchell could unstick in the very short distance of 700-750ft (213-229m) and the wind over the *Hornet's* deck would reduce that further.

The presence of a Japanese patrol boat dictated an earlier than planned departure for the Mitchells on the day of the raid, when they were still 800 miles (1,300km) from Tokyo. All 16 Mitchells survived the run in and the raids over Japan but were subsequently lost due to bad weather and a shortage of fuel while attempting to find their designated land-

The definitive Mitchell, a B-25J complete with forward mounted dorsal turret, 'cheek' forward fuselage guns, waist guns, twin tail guns and no ventral turret. B-25J production amounted to 4,318 in 1944 and 1945, nearly half of the Mitchell total.

ing grounds in China. One put down near Vladivostok in the Soviet Union where both the crew and the aircraft were interned. The Soviets had seen the Mitchell before as delivery of the first of an eventual 870 supplied under Lend-Lease had begun the previous month. Most of the crews were saved although some were captured by the Japanese and two of them executed in a vindictive act which was unfortunately typical of Japan's treatment of many prisoners of war.

For his part in leading the raid, Jimmy Doolittle was awarded the Medal of Honour, while Lt Ted Lawson, the skipper of one of the Mitchells, wrote the book *30 Seconds Over Tokyo* (a best seller during the war) from which the well known film of the same name starring Spencer Tracy was made.

In Ever Larger Numbers

By the time deliveries of the further improved B-25C and B-25D (company designations NA-82 and NA-87) began in early 1942, the Mitchell was subject to large scale production in several factories. The original factory at Inglewood had witnessed all B-25 production up to this point and the manufacture of 1,619 B-25Cs continued there. The B-25D was the same aircraft manufactured at a new facility in Kansas City, 2,290 of them being turned out. In 1944, all Mitchell production was transferred to Kansas City while Inglewood was devoted exclusively to the P-51D Mustang along with another factory in Dallas, Texas.

The B-25C and D featured numerous changes when compared to the B-25B. Powered by 1,700hp R-2600-13 engines and fitted with an autopilot (which had first been used on the Tokyo raid aircraft), the new models had a normal maximum takeoff weight of 29,500lb (13,381kg) and a rarely used maximum overload weight of no less than 41,800lb (18,960kg) was permitted under special circumstances.

Defensive armament remained as before but the normal main wing fuel tanks could be supplemented by a 487imp gal (2,214 l) bomb bay tank.

When this tank was not fitted, alternative bomb loads of one 2,000lb (907kg), two 1,000lb (454kg), six 500lb (227kg), eight 250lb (113kg) or twelve 100lb (45kg) bombs could be accommodated internally.

As the B-25C and D were developed, further wing fuel capacity was incorporated and that combined with a waist tank and the bomb bay tank meant a Mitchell could uplift a total of 1,402imp gal (6,373 l) of fuel, in which case the takeoff weight was increased to 33,500lb (15,195kg).

The armament – both offensive and defensive – was also increased as the B-25C/D evolved. An extra gun was mounted in the nose position and in addition to the bomb combinations listed above, it became possible to carry two 1,600lb (726kg) armour piercing or three 1,000lb (454kg) bombs internally. With the addition of eight external racks, a further 2,000lb (907kg) could be carried bringing the maximum permissible weight up to 5,200lb (2,359kg) over short ranges. It was also possible to carry a 2,000lb (907kg) torpedo under the fuselage.

The B-25C and D served with several foreign air arms including the RAF (335 supplied as the Mitchell II), Canada, Brazil and the Netherlands government, some of which served with No 18 Squadron NEIAF in Australia. Of these, 30 were then passed on to the RAAF.

The US Navy also operated these Mitchell variants as the PBJ-1C and -1D equipped with radar and depth charges and ten B-25Ds were converted to unarmed photographic-reconnaissance configuration for the USAAF in 1943 under the designation F-10.

The XB-25E and F were one-off experimental variants converted from B-25C airframes for the testing of hot air and thermal de-icing systems for the wings and tail surfaces.

Mitchell Big Hitter

Although the Mitchell was and would remain primarily a medium bomber, its versatility in other roles was a continuing asset. Among those other roles was ground strafing, and several aircraft had been modified in the field to accept a battery of usually eight 0.50in machine guns in the nose. The use of a much bigger gun for air-to-ground duties had been investigated in the USA and it was discovered that a standard US Army M4 75mm field gun could be installed in a Mitchell.

The production application of this installation was called the B-25G, 405 examples of which were produced in 1942. The 9ft 6in (2.89m) long, 900lb

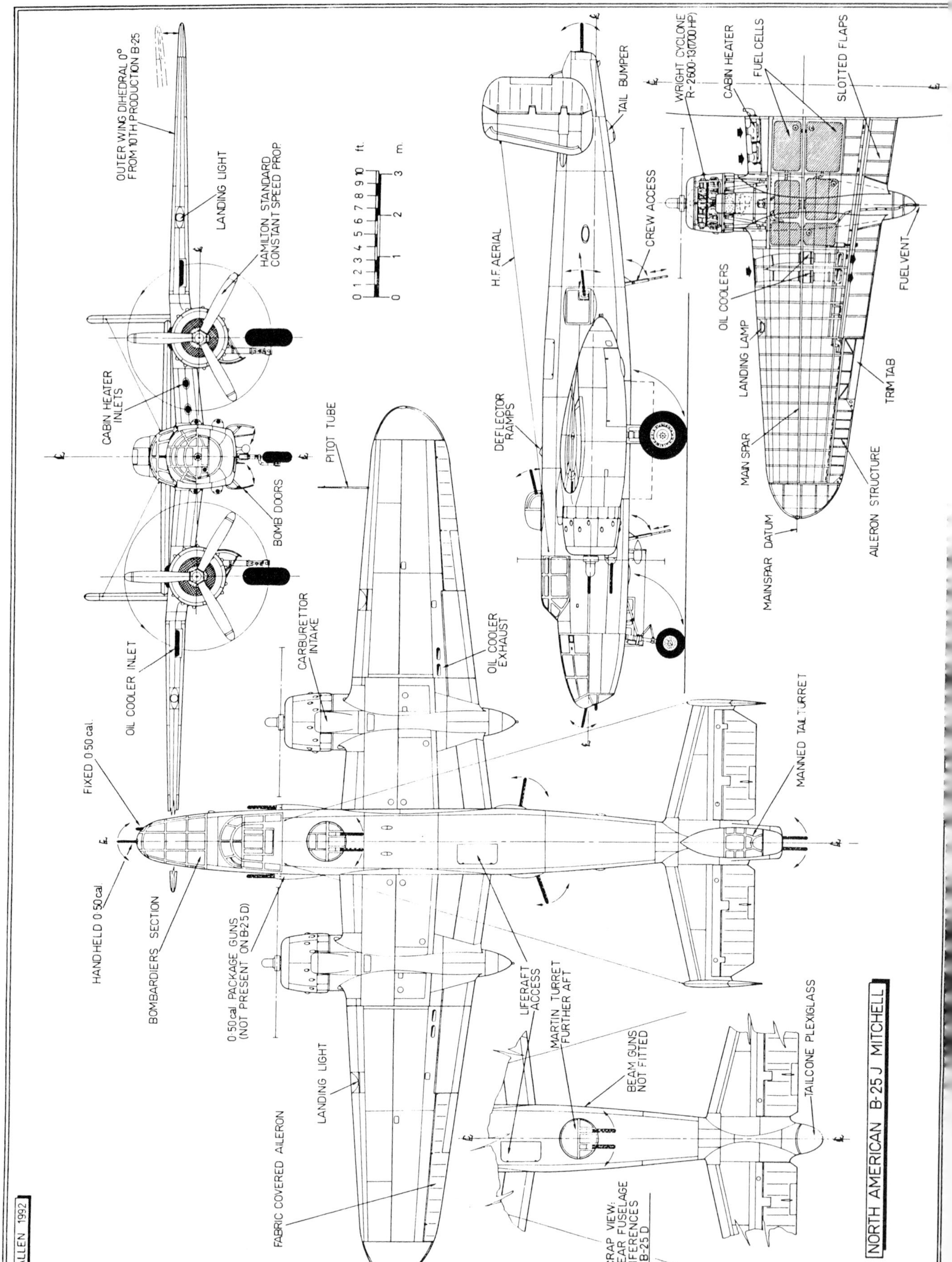
NORTH AMERICAN B-25 J MITCHELL
© J F ALLEN 1992
OUTER WING DIHEDRAL 0° FROM 10TH PRODUCTION B-25
LANDING LIGHT
HAMILTON STANDARD CONSTANT SPEED PROP.
CABIN HEATER INLETS
BOMB DOORS
OIL COOLER INLET
FIXED 0·50 cal.
HANDHELD 0·50cal.
BOMBARDIERS SECTION
0·50 cal. PACKAGE GUNS (NOT PRESENT ON B-25 D)
LANDING LIGHT
FABRIC COVERED AILERON
PITOT TUBE
CARBURETTOR INTAKE
OIL COOLER EXHAUST
LIFERAFT ACCESS
MARTIN TURRET FURTHER AFT
BEAM GUNS NOT FITTED
*SCRAP VIEW: REAR FUSELAGE DIFERENCES B-25 D
TAILCONE PLEXIGLASS
MANNED TAIL TURRET
0 1 2 3 4 5 6 7 8 9 10 ft.
0 1 2 3 m.
H.F. AERIAL
DEFLECTOR RAMPS
CREW ACCESS
TAIL BUMPER
WRIGHT CYCLONE R-2600-13 (1700 HP)
CABIN HEATER
FUEL CELLS
SLOTTED FLAPS
OIL COOLERS
LANDING LAMP
MAIN SPAR
MAINSPAR DATUM
AILERON STRUCTURE
TRIM TAB
FUEL VENT

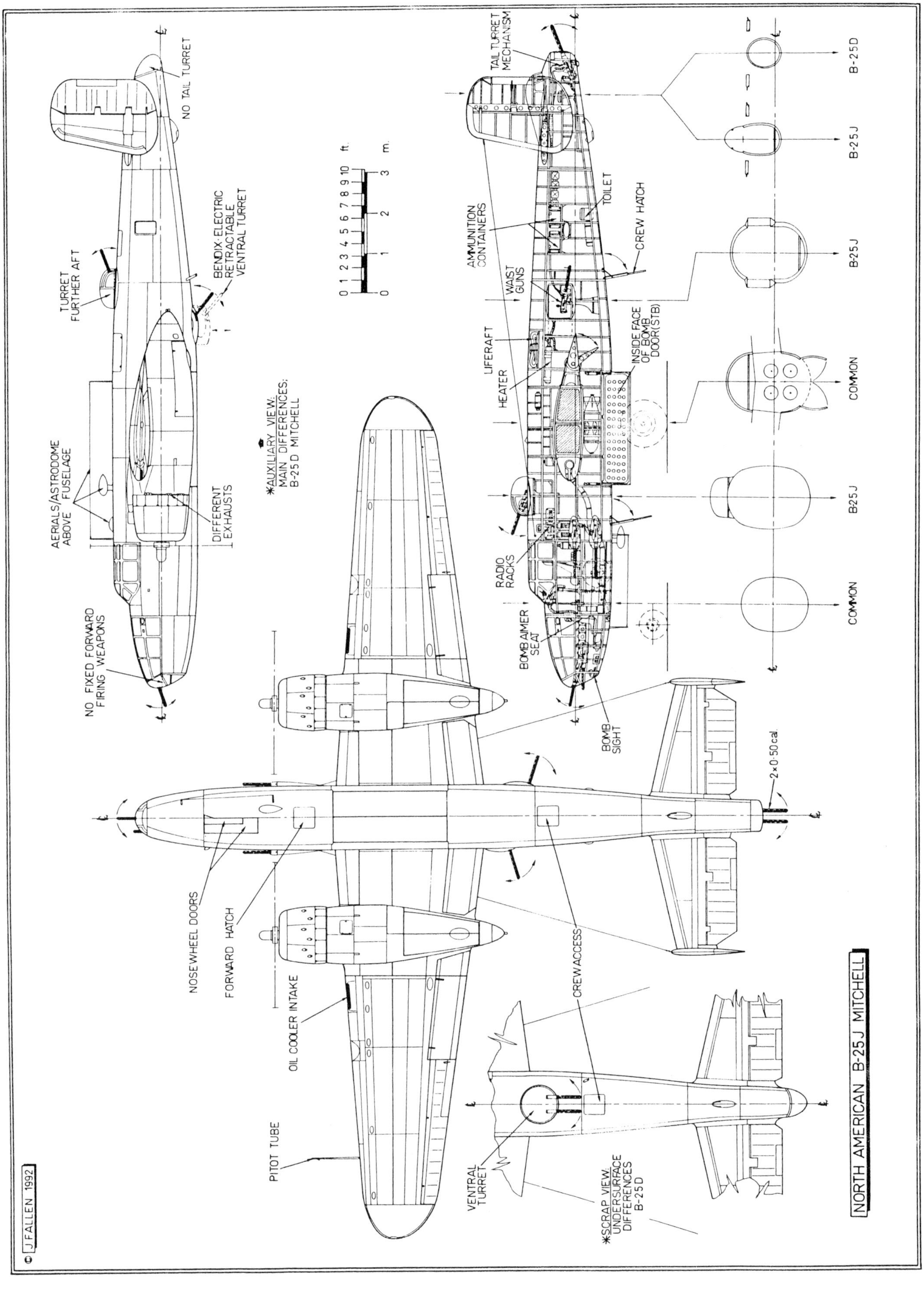
NORTH AMERICAN B-25J MITCHELL
© J F ALLEN 1992
NO TAIL TURRET
TURRET FURTHER AFT
BENDIX ELECTRIC RETRACTABLE VENTRAL TURRET
AERIALS/ASTRODOME ABOVE FUSELAGE
DIFFERENT EXHAUSTS
NO FIXED FORWARD FIRING WEAPONS
*AUXILIARY VIEW: MAIN DIFFERENCES: B-25D MITCHELL
0 1 2 3 4 5 6 7 8 9 10 ft.
0 1 2 3 m.
TAIL TURRET MECHANISM
AMMUNITION CONTAINERS
TOILET
CREW HATCH
WAIST GUNS
INSIDE FACE OF BOMB DOOR(STB)
LIFERAFT
HEATER
RADIO RACKS
BOMBAIMER SEAT
BOMB SIGHT
B-25D
B-25J
B-25J
COMMON
B-25J
COMMON
2×0·50cal.
NOSEWHEEL DOORS
FORWARD HATCH
OIL COOLER INTAKE
PITOT TUBE
CREW ACCESS
VENTRAL TURRET
*SCRAP VIEW: UNDERSURFACE DIFFERENCES B-25D

As was the case with many other Allied aircraft, the Soviet Union was a major recipient of the Mitchell, some 870 being supplied under Lend-Lease of which more than 300 were B-25Js as illustrated here.

(408kg) gun was mounted in a cradle which extended under the pilot's seat and projected forward into an armoured and shortened nose assembly. A hydro spring mechanism was incorporated to take the 21 inch (53cm) firing recoil and the aircraft carried 21 of the gun's 15lb (7kg) shells which were individually hand loaded by the navigator/gunner. Two 0.50in machine guns were also mounted in the nose for sighting and as anti-flak weapons.

The crew was reduced from five to four – pilot, navigator/gunner, dorsal turret gunner and radio operator/ventral gunner, while the modification did cost the aircraft about 10mph in maximum speed which was reduced to about 275mph (442km/h).

Although fundamentally successful, the B-25G was far from perfect in operations. Usually only about four of the 75mm gun's rounds could be fired in a single pass and the aircraft was vulnerable to ground fire as it had to hold a steady dive during the attack. Nevertheless, the Mitchell proved to be a solid and stable platform for such an installation and further development would follow.

The concept of the Mitchell 'gunship'

A snowscape surrounds RAF B-25J Mitchell III HD397 of 180 Squadron during service in Germany at the end of the European war. Note the lack of side forward fuselage guns. (via Neil Mackenzie)

was continued with the B-25H (NA-98), the first example of which flew in July 1943. Production reached exactly one thousand.

Compared with the B-25G, the new model featured several improvements, not the least of which was the use of a different and notably lighter 75mm nose gun, the T13E1. Supplementing that was an array of weapons which gave the B-25H an extraordinary firepower: four 0.50in machine guns in the armoured and truncated nose, two further 50 calibres on each side of the forward fuselage, twin 'fifties' in a new dorsal turret which was moved forward to a point just aft of the cockpit, two more in the tail position and yet another in waist hatches – a total of fourteen machine guns and the 75mm artillery piece.

In addition, this Mitchell could also carry a bomb load of 3,200lb (1,451kg), while the crew was restored to five – pilot, navigator/radio operator/gunner, flight engineer/dorsal gunner, midships gunner/camera operator and tail gunner. Powerplants were the R-2600-13 or -29, still rated at 1,700hp. Despite being an improvement on the previous model, the B-25H still suffered most of the disadvantages of its predecessor, namely slow rate of fire from the 75mm gun and vulnerability to unfriendly fire during its attack run. For these reasons, it had been largely removed from service by August 1944, just six months after having been first introduced to operational units in the Pacific.

The Definitive J

By far the most numerically important Mitchell variant was the B-25J (NA-108), no fewer than 4,318 of which were built in 1944 and 1945. The last Mitchell to be manufactured – the 9,816th – was a B-25J delivered in August 1945.

With the B-25J, the Mitchell returned to its original concept of medium bomber, although aft of the nose the airframe and armament (including the four side fuselage guns) was similar to that of the B-25H. Gone was the 75mm gun and the four nose mounted 'fifties', replaced with a B-25C/D style glazed nose for the navigator/bombardier complete with a single flexibly mounted and two fixed 0.50in machine guns. The R-2600-29 engine was installed, rated at 1,700hp for normal takeoff and 1,850hp war emergency and the maximum bomb load was 4,000lb (1,814kg) for short ranges although 3,000lb (1,361kg) was the norm – three 1,000 pounders or six 500 pounders.

The RAF received 314 B-25Js as the Mitchell III, the Soviets received a similar number, while 20 of the 50 Mitchells which found their way into RAAF service were of this model.

Many B-25Js operated by the USAAF in Europe were modified for the ground attack role by the installation of eight 0.50in machine guns in a 'solid' nose. In addition, some 800 Js were built new to that specification.

Other Mitchell variants were conversions of the major production variants and included the RB-25 transports which were used by USAAF chief General H H (Hap) Arnold and General Dwight Eisenhower and others, the F-10 photo-reconnaissance conversions, the US Navy's PBJ-1 series (706 delivered in several versions) and the AT-25 series of mainly bombardier trainers. Postwar, these and other Mitchells converted for training duties were redesignated TB-25s. These served with the United States Air Force until well into the 1950s, while Canada had some on strength in the early 1960s. Some Central and South American air arms used Mitchells operationally during the two decades following the end of World War II in local squabbles, but the aircraft's combat career proper ended in 1945 with the capitulation of Germany and Japan.

NORTH AMERICAN B-25J MITCHELL III
Powerplants: *Two Wright R-2600-29 Double Cyclone 14 cylinder two row air cooled radial piston engines with two speed mechanical superchargers, each rated at 1,700hp for takeoff, 1,850hp war emergency and 1,450hp at 13,000ft. Three bladed Hamilton Standard constant speed fully feathering propellers. Fuel capacity 812imp gal (3,691 l) in four main and two groups of three auxiliary tanks in wings plus a fixed tank at the top of the bomb bay of 179imp gal (814 l) capacity; droppable 279imp gal (1,268 l) ferry tank in bomb bay.*
Dimensions: *Wing span 67ft 7in (20.60m); length 52ft 11in (16.13m); height 15ft 10in (4.82m); wing area 610sq ft (56.67sq m).*
Weights: *Empty 21,100lb (9,571kg); normal loaded 33,500lb (15,196kg); max overload 41,800lb (18,960kg); loadings (at 33,500lb) power 9.85lb/hp, wing 54.92lb/sq ft.*
Armament: *Maximum bomb load 4,000lb (1,814kg), normal 3,000lb (1,361kg); total of 13 0.50in Colt-Browning machine guns (bomber version) with 5,000 rounds of ammunition comprising one fixed and two movable guns in nose, four fixed on fuselage sides, two in dorsal turret, two in tail and two in waist; ground attack version with eight 0.50in guns in nose; some with underwing stub mountings for four 5 inch rocket projectiles.*
Performance: *Max speed 275mph (442km/h) at 15,000ft; cruising speed 230mph (370km/h); initial climb 1,110ft (338m)/min; service ceiling 24,000ft (7,315m); range (3,200lb bombs, wing tanks full) 1,275 miles (2,050km), ferry range 2,700 miles (4,345km).*

More than 120 Mitchells survive in the 1990s, including many which are airworthy. Among them is this B-25H (ex 44-4432) now operated by the EAA Aviation Foundation with US civil registration N10V. The aircraft has a B-25J nose section and is one of the Mitchells which flew in the 1969 film 'Catch 22'. (Jim Thorn)

MITCHELLS IN AUSTRALIA

The Dutch Connection

Although the Royal Australian Air Force did not operate B-25 Mitchells in its own right until 1944 – and then by a single squadron – arguably the most successful medium bomber of the war did enjoy a very close relationship with the RAAF for some two years before that due to its association with the Netherlands East Indies Air Force's No 18 Squadron.

Indeed, of the 50 Mitchells which carried RAAF markings and serial numbers (30 B-25Ds and 20 B-25Js), no fewer than 39 had either served with or had been allocated to No 18 Squadron NEI before the RAAF took them over.

The Dutch presence in Australia resulted from the Japanese invasion of the Netherlands East Indies (now Indonesia) in early 1942. Dutch airmen who had survived the bitter fighting there or were already in Australia and Ceylon to pick up and deliver Mitchells ordered for the NEI formed the nucleus of the Netherlands East Indies Air Force in Australia.

Two NEI combat squadrons were formed as a result of this: No 18 with its Mitchells and No 120 (Fighter) Squadron with Curtiss P-40 Kittyhawks. A third squadron – No 119 – existed only on paper and very briefly in late 1943. It was intended to be equipped with Mitchells and act as a training/reserve unit.

Shortly before Japan entered World War II, the Netherlands East Indies Government had placed substantial orders for combat aircraft as the situation grew more threatening. Among them were orders covering 162 B-25C Mitchells.

As Mitchell production was running behind schedule in late 1941 and the situation in the Netherlands East Indies was becoming more urgent every day, B-25Cs intended for the USAAF were allocated to the Dutch order – the queue was 'jumped' – but this did not prevent the first aircraft which were ready for despatch in February 1942 being too late to help defend the NEI against the Japanese. By then, the invaders were already overrunning the area; by early March it would be all over.

The original delivery plan allowed for the B-25Cs to be ferried to the NEI by Consolidated Aircraft crews, interestingly, working under contract to the Dutch and by the Royal Air Force's Ferry Command. The situation in the NEI by February 1942 necessitated a hasty revision of these plans including the delivery of the first eight aircraft to India from the USA. Of the eight, two were damaged in the USA and remained there and a third crashed in India *en route*. The five survivors were requisitioned by the RAF and modified to photo-reconnaissance configuration.

In the meantime, 18 Dutch crews were waiting at Brisbane's Archerfield aerodrome in anticipation of receiving their new Mitchells, the first of which began arriving in early March. Discussions between the Commander-in-Chief Netherlands East Indies Forces in Australia (Maj Gen L H Van Oyen) and the Australian Chief of Air Staff (ACM Sir Charles Burnett) resulted in agreement to form a Dutch squadron comprising 18 Mitchells in Australia. Initially at least, the squadron would form part of the RAAF.

A letter from Burnett to Van Oyen outlined the salient points of the agreement: "... With regard to the organisation of the squadron, it is noted that the NEI organisation differs somewhat to the RAAF and in this connection it is suggested that while the squadron is operating in Australia it would simplify operational planning and control if it were organised on the same basis as our own.

"The RAAF squadron organisation provides for three flights of four aircraft and six reserve aircraft in the Maintenance Flight. In the matter of aircrews our policy provides for a crew for both first-line and all reserve aircraft, plus an additional reserve of three complete crews. This is considered necessary for operations in Australia and provides for all aircraft of the squadron being moved to a distant location without delay and also a reserve against war wastage.

"RAAF personnel will be provided to fill the posts in the squadron establishment not filled by NEI personnel, and also any additional personnel experience shows to be necessary for the efficient operation of the squadron. It is intended to provide an officer in the unit who will be responsible for any administration and discipline of RAAF staffs in the unit ..."

A well-known publicity shot of RAAF B-25J Mitchell A47-44. This aircraft was delivered in April 1945 and disposed of in 1950. (RAAF)

B-25C Mitchells of No 18 Squadron Netherlands East Indies photographed in 1942 whilst based at Canberra. N5-132/41-12919 in the foreground crashed shortly after takeoff from MacDonald NT in February 1943. (via L E Benoit/Mike Kerr)

The question of the precise status of the squadron remained an issue until early July 1942 when it was decided that it would no longer be an RAAF but a Dutch unit, albeit still operating under the command of the RAAF and containing a healthy proportion of RAAF personnel. The change caused some administrative problems but it meant that the squadron could now be organised and trained to NEI standards and under Dutch supervision.

Mitchells Arrive

It took rather longer than anticipated for No 18 Squadron to receive its planned complement of Mitchells. In March 1942 the Archerfield detachment had received 12 of the promised 18 B-25Cs but these were immediately and unofficially "purchased" by the USAAF 5th Air Force to help in the desperate New Guinea campaign, where the Japanese were advancing strongly towards the vitally important town of Port Moresby.

This arrangement caused chaos among the record keepers and those involved in the accountancy of the Dutch Mitchells as the Netherlands Purchasing Commission had been charged for the

A mixture of national markings on B-25C Mitchells prior to delivery from the USA including (second from bottom) Dutch aircraft N5-126/41-12462. This aircraft was part of the second batch of five Mitchells which arrived in Australia in June/July 1942.

Newly delivered B-25C and B-25D Mitchells photographed in late 1942 , shortly before No 18 NEI Squadron moved to the Northern Territory.

aircraft. The next batch of Mitchells were promised to the Dutch, but these, too, were taken over by the USAAF.

No 18 Squadron was formed at Canberra on 4 April 1942 under the command of Lt Col Fiedeldij and as an RAAF, not Dutch, unit. Finally, in the same month, the squadron was given five B-25C Mitchells. They carried the Dutch serial numbers N5-132, 134, 135, 136 and 151. In late June/early July another five (N5-122, 124, 125, 126 and 127) were taken on strength, partially as replacements for the first five aircraft, some or all of which apparently disappeared into USAAF service at around this time.

The word 'apparently' is used as the records are unclear. Certainly, Mitchells came and went during the first months of 18 Squadron's Australian existence, because when 'regular' deliveries began in late August 1942 (18 B-25Cs and Ds) several of the original serial numbers were duplicated on the new aircraft. It is known that N5-123, 124, 125 and 127 were handed over to the USAAF in September and that N5-122 followed them in December. N5-126 had meanwhile become the squadron's first casualty. In July it had made a precautionary landing at Moruya NSW, during which the main undercarriage collapsed. The aircraft was eventually repaired and returned to the USAAF.

18 Squadron's first commanding officer, Lt Col B J Fiedeldij.

If No 18 Squadron had to be patient to get its Mitchells in the first place, it also had to fight to keep them. By mid June 1942 the squadron's strength comprised five Mitchells and 15 newly delivered Douglas Bostons, ten of which were DB-7B aircraft originally on the books of No 22 Squadron RAAF and the remainder USAAF A-20As.

The Bostons were supposedly "in exchange" for the Mitchells, and on 13 June nine USAAF pilots arrived at Canberra to take away the Mitchells, the intention being to ferry them to Charters Towers (Qld) and thence on to active service with the USAAF 5th Air Force.

But 18 Squadron's Lt Col Fiedeldij knew nothing of this, and apart from anything else, wanted to keep the Mitchells as they were regarded as being superior

Mitchell N5-151 flown by Lts Winckel and Van Loggen scored 18 Squadron's first claimed combat success when it sunk a Japanese submarine off Moruya NSW in June 1942. (via L E Benoit/Mike Kerr)

The RAAF received the first of 30 B-25D Mitchells in April 1944, all of them transferred to No 2 Squadron from Dutch stocks. This example is interesting as it is one of several hybrids which featured some B-25J characteristics including the waist guns. (via Neil McKenzie)

to the Bostons. As a result, he refused to hand the Mitchells over to the Americans and it has been reported that he went as far as locking some of them up!

A short stalemate developed and on the following day orders were issued by NEI Headquarters to retain both types for the moment. On 16 June the first Bostons were taken away – by the end of the month all had gone including the ten RAAF DB-7Bs which were returned to No 22 Squadron.

It must be remembered that as far as the Dutch were concerned they had ordered 162 Mitchells and expected them all to be delivered. They were, but the last of them was handed over just before the war ended.

Moving North and More Mitchells

No 18 Squadron's activities during its stay in Canberra revolved around training and anti-submarine patrols, one of the latter claiming the sinking of a Japanese submarine off the coast of New South Wales on 5 June. The Mitchell involved was N5-151, flown by Lts Winckel and van Loggen.

The action undoubtedly elevated the status of the squadron, which was by now keen to move north to carry on the fight against Japan. A shortage of aircraft generally on Australia's east coast and 18 Squadron's own lack of aircraft precluded this for the moment, but with the arrival of the 18 'permanent' Mitchells in August and September, a move came closer to reality.

The lack of some items of equipment concerned the senior Dutch officers. The Mitchells were fitted with unfamiliar D8 bombsights instead of the preferred Norden sights and despite the fittings for long range wing fuel tanks being installed in the aircraft, the tanks themselves were nowhere to be seen. Equipment shortages were a general problem at the time, and as no Norden bombsights were available, the Americans offered the services of an instructor on the D8 sight.

In October 1942 the squadron was readied for the move north to its new base at MacDonald in the Northern Territory. An advance party left Canberra for the Territory in late November, followed by the main groups of equipment, personnel and aircraft during December. The Mitchells were slightly delayed due to the need to fit them with long range fuel tanks in the bomb bays. This task was performed by No 1 Aircraft Depot at Laverton, Vic, at the rate of one per day and the Mitchells headed north in the last few days of 1942.

No 18 Squadron's active service is described in the following chapter.

Discounting the ten 'temporary' B-25s delivered to 18 Squadron between March and July 1942, a total of 150 Mitchells were taken on strength by the NEIAF, 19 in 1942, 16 in 1943, 87 in 1944

A group of 18 Squadron personnel photographed in 1942 illustrates the mix of Dutchmen and Australians which made up the squadron. From left: Wnt Off Hutchison (RAAF), Capt Tander (NEI), Lt Hagers (NEI), Sgt Robinson (RAAF), Adj Verblaak (NEI), Sgt Maj Stellema (NEI), Adj Bogels (NEI) and Lt Olsen (NEI).

Twenty B-25J Mitchells entered RAAF service. A47-30 was one of the first batch of seven taken over from Dutch stocks in June, July and August 1944.

and 28 in 1945. Following the delivery of the batch of 18 aircraft in August/September 1942, no more were received until April 1943 when a further eight arrived.

Larger numbers were received in 1944 including some 50 in the first three months, partially because No 18 Squadron was also allotted the aircraft intended for the still-born No 119 Squadron, among them the first B-25Js from mid 1944. The number of aircraft available was therefore far greater than those needed for a single squadron, so arrangements were made to re-equip the RAAF's No 2 Squadron with the excess aircraft.

RAAF Mitchells

No 2 Squadron RAAF could trace its origins back to 1916 when its predecessor was formed as a unit of the Australian Flying Corps in Egypt. Renamed No 2 Squadron in 1918 and disbanded the following year after the cessation of hostilities, the squadron remained in limbo until May 1937 when it was reformed as No 2 (General Reconnaissance) Squadron at Laverton, Victoria, equipped with Hawker Demons.

RAAF MITCHELLS SUMMARY OF DELIVERIES

Serial No	Model	No	Delivery
A47-1/20	B-25D	20	04/44
A47-21/25	B-25D	5	06/44
A47-26/32	B-25J	7	06-08/44
A47-33/37	B-25D	5	08/44
A47-38/39	B-25J	2	09/44
A47-40/48	B-25J	9	04-06/45
A47-49-50	B-25J	2	08/45

Avro Ansons comprised the squadron's main strength in the early months of World War II until these were replaced by Lockheed Hudsons in June 1940. The Hudson remained No 2 Squadron's primary equipment until January 1944 when Beauforts replaced the American aircraft, but these in turn were replaced by Mitchells just four months later. In the meantime, the squadron had seen active service against Japan in the northern Australia/Timor/NEI/Borneo regions, conducting maritime patrol, bombing, reconnaissance and anti-shipping missions.

At the time of changeover from Beauforts to Mitchells, No 2 Squadron was based at Hughes in the Northern Territory under the command of Wng Cdr L A Ingram. Together with No 18 (NEI) Squadron based at Batchelor to the south, No 2 Squadron formed No 79 Wing RAAF.

The first batch of 20 B-25D Mitchells were transferred from the Dutch to the RAAF in late April 1944. Serial numbers were A47-1 to A47-20) and these were followed in June and July by five more B-25Ds (A47-21 to 25) and seven B-25Js (A47-26 to 32). The remaining Mitchells delivered to the RAAF from Dutch allotments were five B-25Ds (A47-33 to 37) in August 1944 and two more B-25Js (A47-38 and 39) in September.

The last batch of five B-25Ds was interesting as at least three of them (A47-34, 36 and 37) are known to have been hybrids which featured some of the characteristics of the later B-25J including waist gun positions below the mid-upper turret and 'cheek' gun packs on the lower forward fuselage in line with engine nacelles. These modifications were applied to several Dutch B-25Ds.

The transfer of the initial batch of Mitchells from NEI to RAAF stocks was accomplished at Canberra in April 1944 and with few hours on their airframes, were in generally good condition. Required modifications were minimal and included the fitting of bomb bay fuel tanks and replacement of Norden bomb sights by Mk.XI units.

With the completion of this and other maintenance work, the Mitchells were taken to the Northern Territory to join No 2 Squadron.

The final 11 Mitchells delivered to the RAAF were B-25Js A47-40 to 50 and were handed over between April and August 1945. These aircraft did not come from Dutch stocks but direct from the USAAF and only two of them were taken on strength by No 2 Squadron.

A lineup of No 2 Squadron B-25D and B-25J Mitchells at Hughes in the Northern Territory. (via Neil Mackenzie)

18 Squadron NEIAF

18 Squadron's move to MacDonald Strip in the Northern Territory was not an easy one with transport difficulties and the discovery on arrival that there was still much work to be done on both the aerodrome itself and the camp site. The airstrip was made of granite gravel and although serviceable, was easily damaged by rain and required constant maintenance. There was no fresh water on the base, insufficient huts and buildings and taxiways and dispersals had to be prepared. Conditions remained relatively primitive for all involved at MacDonald.

The personnel make up of the squadron at the beginning of 1943 comprised 253 NEI and 287 RAAF officers and men.

No 18 (NEI) Squadron officially completed its move to the Northern Territory on 18 January 1943 and came under the command of RAAF North Western Area. With it in the Territory in the first half of 1943 were six Australian, one American and one British squadron. The Australian squadrons were No 2 (Hudsons based at Hughes), No 12 (Vengeances, Batchelor), No 13 (Hudsons, Hughes), No 31 (Beaufighters, Coomalie Creek), No 452 (Spitfires, Strauss) and No 457 (Spitfires, Livingston). The USAAF's 319th Squadron (Liberators) was based at Fenton and the Spitfires of No 54 Squadron RAF at Winnellie, Darwin.

Apart from Winnellie, all these bases were south/south-east of Darwin with MacDonald, about 170 kilometres away as the crow flies, the farthest distant. The town of Katherine was a further 100 kilometres away.

The basic task of the aircraft based in the area was to defend Australia's northern coastline and the Torres Strait area. For the fighter squadrons it meant meeting enemy air raids, but for the bomber and strike squadrons it was a case of attack being the best form of defence, with enemy airfields and shipping to the north of Australia coming in for constant attention from the Liberators, Beaufighters and Mitchells. Although the number and intensity of Japanese raids generally diminished as 1943 progressed, fighter opposition over targets in Ambon, the Arafura Sea islands and Timor remained strong.

18 Squadron completed its first operational mission on 18 January – a reconnaissance to Tanimbar Island in the Arafura Sea directly north of Darwin. The object of the exercise was to find some enemy shipping which had been reported, without success.

The relative peace of that mission was short lived. On the following day a raid over Timor resulted in some damage to the Mitchells from enemy fires but on the 20th, 18 Squadron claimed its first 'kills' with the probable destruction of two Mitsubishi 'Zeke' fighters and one Nakajima 'Dave' seaplane during combat over the Japanese airbase at Fuiloro on Timor.

With increasing numbers of raids on Japanese installations and airfields and masthead height attacks on shipping came the inevitable casualties. Several aircraft suffered damage from their own bombs' explosions in the latter sorties, while at the end of January, two Mitchells (N5-134 and N5-139) had to make forced landings when they found themselves running out of fuel on the way back from a raid on Dili, a raid by five Mitchells during which no bombs were dropped thanks to heavy rain and poor visibility over the target. Happily, the Mitchells came down on Australian soil with a minimum of drama, N5-134 on the runway at Port Keats (on the Northern Territory's

Two Of No 18 (NEI) Squadron's Mitchells photographed in late 1942 or early 1943, about the time the squadron moved to the Northern Territory. N5-139 crashed landed at Port Keats in February 1943 and N5-140 was lost on operations the following April. (via Neil Mackenzie)

With a typical tropical cloudscape in the background, two of No 2 Squadron's Mitchells are photographed from another squadron aircraft. The date is November 1944, a time when the Mitchells of No 79 Wing were enjoying considerable success in the anti-shipping role. (via Neil Mackenzie)

N5-146, the aircraft of 18 Squadron's Lt 'Gus' Hagers. The Mitchell was named 'Lienke' after Hager's wife, who was a prisoner of the Japanese. (via L E Benoit/Mike Kerr)

As mentioned in the previous chapter, some of 18 Squadron's early Mitchells had their serial numbers duplicated later on. This is the second N5-144, delivered in September 1942 and lost during a raid on Dili in February 1943. This was the first Dutch Mitchell to be lost in action.

western coast near the northern extremity of the Western Australia border) and N5-139 on marshy ground.

It wasn't long before the first tragic loss occurred. On 5 February N5-132 crashed near MacDonald killing all nine on board, including three RAAF personnel. The aircraft crashed shortly after an early morning takeoff from MacDonald, the load of bombs and fuel combining to cause a large explosion which was clearly heard by everyone at the base, some six kilometres away.

The first 18 Squadron Mitchell to be lost in action was N5-144 on 18 February. During a raid on Dili in Timor, the six attacking Mitchells were intercepted and then forced to engage in a running battle with two 'Zekes'. The result was one of the Japanese fighters probably destroyed but at the cost of N5-144 which went down in the sea.

Its pilot, Lt B Grummels and bombardier Sgt R G Tyler, were both killed but the remainder of the crew took to its dinghy and was picked up by a Royal Australian Navy ship after first having been dropped supplies from RAAF Hudsons. All had been wounded. During the ordeal on the dinghy, the crew had been subject to serious harassment by a large shark!

18 Squadron's Intelligence Officer issued the following report on the demise of N5-144, based on what the survivors had told him. It well illustrates that war is not a pleasant thing: "... within a couple of minutes oil pressure dropped to zero and the port engine stopped. Almost immediately the Zeke made a pass at them from above, and the pilot was killed by a M/G bullet entering the top of his head, then coming out and going into his right

Rabaul was a regular target for Allied aircraft. This shot was taken during a raid by Mitchells in November 1943. (RAAF)

N5-138, one of the early batch of Mitchells delivered to 18 Squadron in 1942. (via Neil Mackenzie)

A pair of 18 Squadron Mitchells low over the Northern Territory, photographed from a third aircraft. Note the 'strafer' noses on these aircraft, fitted to increase their effectiveness in the anti-shipping role. (via Neil Mackenzie)

18 Squadron's N5-145 'The Flying Dutchman' ended its operational career with this mishap at Batchelor in October 1943. The pilot attempted a downwind landing, with familiar results. The Mitchell was converted to components.

knee. The bombardier in the nose of the aircraft was probably killed by the same burst.

"The co-pilot took over control of the aircraft and called for help from the other five B-25s. The same Zeke made three more overhead attacks. The pilot watched until seeing him roll over and then turned into the attacker each time who seeing the guns pointing his way broke off. The Zeke then made a fourth attack ... hit the starboard engine aileron and rudder which tore off.

"By this time the rest of the B-25s had seen and heard what was happening and returned to help and drove off this Zeke which returned towards Timor. N5-144 then flew on for 20 miles on one engine steadily losing power and oil pressure and the aircraft was practically uncontrollable. It went into the sea tail first and bottom turret down – consequently the nose finally went right under.

"The co-pilot and mechanic got out by sliding the side panels by the pilots' seats and got out the rubber dinghy. The aircraft sank in 2 minutes (not 10 as previously reported) giving the co-pilot just time to smash in the top turret perspex with his hands and drag out with the help of the mechanic the top turret gunner and the bottom turret gunner.

"Both of these men had been banged about in the crash landing and were dazed and resisted rescue – had to be forcibly dragged out. The aircraft sank under them as the last man was dragged out and hauled into the dinghy. During the whole of this action the turrets had been inoperative owing to the main hydraulic system going out of order – probably from M/G fire.

"In the dinghy the two rear turrets men were laid in the bottom and the co-pilot and mechanic sat on opposite sides. A shark broke water within 10 minutes and rose across the edge of the dinghy and snapped at the co-pilot's back. He and the mechanic beat the water to frighten it off and then sat in the bottom of the dinghy. A couple of hours later a big sea bird had to be beaten off as well ..."

The survivors were picked up 14 hours after having ditched.

From the Unit History Sheet

The following is an extract from 18 Squadron's daily records sheets and gives some indication of the squadron at work. It covers just the first ten days of February 1943. The first entry – 1 February – is the day after N5-134 and N5-139 force landed due to a lack of fuel after a sortie over Dili. Times are GMT (Zulu) and the time/date notation reads time/day/ time zone, for example 1300/4/Z would mean 1300 hours Zulu on the fourth day of the month.

1/2/43: Petrol, food and supplies flown by Hudson aircraft ... to Port Keats. N5-134 airborne Port Keats 0001/1/Z for Darwin, landing 0057/1/Z. Arrived MacDonald 0343/1/Z. Supplies, including mosquito nets were dropped to N5-139

Ground crews pose with 18 Squadron's N5-166 'Beth' at No 4 RSU (Repair and Salvage Unit) at Pell, close to the squadron's base at Batchelor. (via Neil Mackenzie)

Plenty of activity in this photograph as a Mitchell is 'turned around' between missions. Just about every maintenance and rearming task was performed outdoors. (via Neil Mackenzie)

A No 2 Squadron B-25D Mitchell with some B-25J features incorporated including the waist and tail gun positions. The photograph is dated May 1944, the month joint operations with No 18 (NEI) Squadron began as part of No 79 Wing. (C N Hill)

and a rescue party comprising a Sergeant Fitter and 3 natives departed from Port Keats in a motor launch at 0430/1/Z. Estimated this party will reach plane in 24 hours. N5-137 (Lt Hagers) proceeded to Coomalie carrying CO 18 Squadron to attend conference ... N5-135 (Lt Bal) proceeded to Coomalie and picked up Lt Col Fiedeldij and returned to MacDonald.

2/2/43: N5-128 (Lt V D Schroeff) and N5-135 (Lt Swane) proceeded to Manbulloo, N5-128 for repairs to landing gear.

Flight No 3 (N5-144 Lt Olsen and N5-133 Lt Hulscher) proceeded to Darwin on a standby duty. N5-138 of this flight unable to takeoff owing to engine trouble. Flt No 2 (N5-129 Lt Winckel, N5-137 Lt Hagers, N5-132 S/M Schalk) proceeded to Darwin on standby duty.

3/2/43: No 2 Flight and N5-133 carried out a successful attack on buildings etc in town of Dobo. N5-144 was to have taken part in attack but due to tyre troubles remained at Darwin. N5-144 returned to MacDonald at 0500/4/Z.

4/2/43: Flight No 5 (N5-143 Lt Wetters, N5-140 Lt Grummels, N5-145 S/M Eikelboom) and Flight No 4 (N5-134 Lt Cooke, N5-136 Lt Bal, N5-135 Lt Swane) proceeded to Darwin on standby duty.

5/2/43: Flt No 2 (N5-129 Lt Winckel, N5-137 Lt Hagers, N5-132 S/M Schalk) proceeded to Darwin for standby duty. At approx 2008/4/Z N5-132 crashed in position 3 to 4 miles north west of strip, about one mile to left of the Darwin road. Personnel in the aircraft were S/M Schalk, Sgt Heile, Sgt Janssen (NEI), Sgt MacPherson and Sgt Walton (RAAF) crew; S/M Kessels, Sgt Maarschalkerweert, Sgt Abeleven (Flight Mechanics) and LAC Palamountain (RAAF).

When the search party located the wreckage of the aircraft all personnel were dead with the exception of Sgt Walton who died whilst being conveyed to hospital. Two 500lb bombs in the aircraft had exploded, and the aircraft is regarded as a total wreck.

Flights Nos 4, 5 and 2 carried out a successful attack on the town of Dobo. All aircraft returned safely.

N5-138 (Ens Kuneman captain) proceeded to Manbulloo and returned to MacDonald.

6/2/43: The remains of the six NEI Members and three RAAF Members who lost their lives in crash (N5-132) on 5/2/42 [sic] were interred in the Adelaide River War Cemetery. The cortege departed MacDonald approx 0800 hours [local time] and was accorded honours by all personnel.

7/2/43: N5-130 (Lt Moorrees captain) made a return trip to Manbulloo.

8/2/43: Successful attack carried out on town of Dobo by nine aircraft. Flight Nos 3, 4 and 5. No enemy interception encountered.

Flt No 2 (N5-129 Lt Winckel, N5-137 Lt Hagers, N5-135 Lt Swane) proceeded to Darwin on standby duties.

10/2/43: N5-144 (Lt Grummels captain) made return trip to Manbulloo. Flight No 2 commenced return journey to MacDonald but were recalled ... Aircraft passed over MacDonald but weather and visibility made a landing hazardous and the flight landed at Coomalie Creek, remained overnight, and proceeded to MacDonald in the morning.

Operational Difficulties and Solutions

Operations in the first few months from MacDonald had revealed several problems. The two major ones revolved around the fact that 18 Squadron's Mitchells were operating at the extreme of their range, leaving little fuel reserves for the trip home. It will be recalled that N5-134 and N5-139 had force landed due to a lack of fuel at the end of January and when N5-144 was shot down into the sea the following month, none of the other aircraft participating in that raid were able to remain over the survivors in their lifeboat for the same reason.

Also, very low level attacks on shipping were not proving to be as effective as they should have been due to poor visibility and a lack of aiming capability in bringing guns to bear on ships' anti-aircraft fire.

Solutions to both problems were offered, the question of range by moving to the New Guinea area (with its shorter range missions) and/or fitting a 300 gallon drop tank in lieu of the ventral turret. This tank would replace the normal bomb bay extra tank which robbed that area of half its capacity.

On the question of armament, 18 Squadron's commanding officer, Lt Col Fiedeldij, also recommended additional fixed armament in the nose of the aircraft.

The squadron didn't deploy to the New Guinea area but in early May 1943 it moved slightly further north to Batchelor, about 100 kilometres closer to Darwin. Batchelor was a much better equipped base than MacDonald and the squadron remained there almost until the end of the war.

On the same day that 18 Squadron began operations from Batchelor (7 May),

A general view of Hughes, with No 2 Squadron's Mitchells lined up. (via Neil Mackenzie)

approval to fit extra forward firing guns in the noses of some of the B-25Ds was given. Equipment for the modifications would be provided by the USAAF and the conversions performed at their facility at Eagle Farm airport, Brisbane. Five aircraft were fitted with eight extra 0.50in machine guns, four in the nose and two pairs mounted in blisters on the side of the fuselage. The ventral turret was removed and 300 gallon drop tanks fitted, giving the aircraft an extra 300 miles (480km) range. These aircraft would become masthead height shipping attack specialists with the pilot controlling the release of the bombs.

The squadron's other 15 Mitchells were also modified but to a lesser extent. They received the four blister mounted 'fifties' but not the nose guns and the bombardier retained responsibility for bomb release.

18 Squadron lost two more aircraft before moving to Batchelor, both in April: N5-140 crashed near Darwin *en route* to Timor on a night mission and N5-135 was shot down during a masthead height attack on shipping near the Aroe Islands. The latter incident proved the squadron's concerns about gun armament shortcomings on these sorties. Not that these modifications guaranteed safety on the low level anti-shipping strikes. The missions were inherently dangerous and indeed, just over a year later in June 1944, the then 18 Squadron CO, Lt Col E J G te Roller and his crew was shot down during an attack on ships near Tioor Island. The aircraft was N5-162.

Flight crews of 'A' Flight, No 2 Squadron, photographed at Hughes Field, south of Darwin in May 1944. They are (from left to right); Back row: Flg Off Isbister, Flt Sgt Woodburn, Flt Sgt Hocking, Flt Sgt Pierce, Flg Off Park, Flg Off Simpson, Flg Off Squires, Flt Lt Greenaway, Flt Sgt MacIntosh, Plt Off Avery, Flt Sgt Allen, Flt Sgt Hawthorne, Flg Off Peters, Flt Lt Higgins.
Middle row: Flg Off Wisnieski, Flg Off Slater, Flt Sgt Purcell, Flt Sgt Christie, Plt Off Fletcher, Sqdn Ldr Hayes, Flt Lt Hill, Wnt Off Bomber, Plt Off Preston, Plt Off Millett, Plt Off Hicks, Flg Off Tomlinson.
Front row: Flt Sgt Ryan, Flt Sgt Muir, Flt Sgt McCallum, Flt Sgt Robinson, Plt Off Head, Plt Off Doggett, Flt Sgt Cunningham. (C N Hill)

Ops From Batchelor

No 18 Squadron remained based at Batchelor for the next two years, during which time it flew many strafing and day or night missions against Japanese positions, airfields and shipping in the Timor, Flores and Ceram areas. There was a regular flow of replacement Mitchells from mid 1943, and the transfer of aircraft to No 2 Squadron RAAF began in April 1944.

Replacement crews from a joint Dutch Army-Navy flying school which had been set up at Jackson, Mississippi in the USA began arriving in September 1943, bringing with them new B-25s with heavier nose armament.

The period November 1943 to early January 1944 was a particularly hectic one for the squadron, which was gradually getting more involved in anti-shipping strikes. During this seven week period, the squadron was credited with the destruction or mortal wounding of six Japanese ships totalling 25,545 tons, along with numerous smaller vessels. These raids were often undertaken in the company of RAAF Beaufighters which offered a degree of protection from enemy fighters.

It was during this period that the effectiveness of 18 Squadron's operations in the Timor and surrounding islands areas was recognised by the adoption of an unofficial badge, the "Dutch Cleanser" showing a peasant girl (complete with clogs) brandishing a straw broom.

In March 1944 the squadron was temporarily deployed to Potshot (Learmonth) on the Western Australia coast as a general build up of forces there in anticipation of a possible Japanese attack. This failed to eventuate and by the end of the month the squadron was back at Batchelor and carrying on where it had been left off.

Another temporary move occurred in late July 1944 when 16 of 18 Squadron's Mitchells and their air and ground crews deployed to Truscott on the northernmost part of Western Australia for three days of photo reconnaissance over Selaru in the Arafura Sea and attacks on shipping.

As 1944 progressed, the sortie rate increased, a record 149 being undertaken in June. By this time there were several changes afoot including the arrival of more capable B-25J Mitchells which gradually replaced the earlier C and D models. With the squadron's targets gradually moving further north as the Japanese were pushed back, it was decided to replace the Mitchells with long range Consolidated B-24 Liberators, with Dutch crews travelling to the USA for training.

B-25D Mitchell A47-2 was converted to components following this incident at Hughes on Boxing Day 1944. (via Neil Mackenzie)

Miscellaneous Mitchells of No 2 Squadron (top to bottom): A47-21/KO-L, A47-34/KO-E 'My Favourite', A47-26/KO-K (in a night operations scheme) and A47-16 (or 18)/KO-F. (via Neil Mackenzie)

The introduction of the B-24 would have required some changes to the Dutch-Australian agreement on the establishment of the Dutch squadrons in Australia. Part of this was the fact that an overall limit of 700 Australians had been set for this exercise and the extras required for the Liberator plan would have taken the number over this. With insufficient Dutch personnel available to make up the shortfall, further Australian contributions were necessary. It was decided not to increase the number, so the Liberator proposal came to nought.

It's interesting to note the truly international flavour of the squadron at this time. Dutch and Australian members comprised the majority but of the approximately 600 personnel on strength there were representatives of no fewer than 38 nationalities who spoke 13 languages!

79 Wing RAAF

The re-equipment of No 2 Squadron RAAF with former NEI Mitchells allowed joint operations between the two units from May 1944, with the Australian squadron flying from its base at Hughes, midway between Batchelor and Darwin.

As the combat components of No 79 Wing, the squadrons together and separately continued striking at targets in the Timor general area, with No 18 beginning a successful series of anti-shipping strikes from Truscott on the northern tip of Western Australia in the second half of the year. Operating from Truscott allowed the squadron a little closer to targets in the Netherlands East Indies area.

Remembering that 18 Squadron was a Dutch unit operating within the RAAF, a special mission on 24 September 1944 was of great significance. This was a long distance propaganda sortie over Java, intended to "show the (Dutch) flag" whilst dropping leaflets over internment camps around Batavia and Bandoeng. This and similar flights which followed was of enormous significance to the Dutch prisoners in the camps. The aircraft had an oversized Dutch flag painted on them and leaflets saying "do not despair" were dropped. For good measure, the Japanese occupied Governor-General's residence at Buitenzorg was strafed during the first of these sorties.

No 2 Squadron wasn't quite able to match the sortie rate of No 18 in the first few months after receiving its Mitchells. As a new type – and an 'orphan' in the RAAF at that – there was a perhaps inevitable problem with supplies and a lack of aircraft spares, although a still reasonable sortie rate was maintained and some notable successes were achieved.

No 2 Squadron's first month flying the Mitchell had been spent training and flying search and patrol sorties but in late June it made its first offensive strike with the new aircraft. The target was the airfield at Lautem West on Timor and the flight of nine aircraft was led by Flt Lt C N Hill. Striking in the late afternoon, the first wave of Mitchells fell short of the target

Mitchell A47-30, one of the first batch of B-25Js transferred from Dutch to Australian stocks in June and July 1944. (via Neil Mackenzie)

but most of those following found their mark and all survived the moderate anti-aircraft fire to return safely.

Throughout the second half of 1944 and into 1945, Nos 2 and 18 Squadrons maintained a high level of activity, usually operating as a co-ordinated team but sometimes singly. The following dates and notes give some indication of the jobs being performed by the Mitchells:

29 July 1944: co-ordinated attack by 12 Mitchells from both squadrons on Dutch Timor's Penfui airfield.

31 July: night raid by 13 Mitchells on Maomere targeting installations, shipping and stores.

2 August: co-ordinated attack by 15 Mitchells on Japanese encampment at Mina River Bridge.

6 August: 2 Squadron's first Mitchell loss; A47-6 crashes during bombing practice.

9 August: daylight attack on Langgoer airfield (Kai Island) by 11 Mitchells.

11 August: further night attack on Langgoer airfield by 12 Mitchells.

17 August: strike against shipping and other enemy activity at Tioor Island by 12 Mitchells. Langgoer airfield made unserviceable by further raid by eight Mitchells. 18 Squadron's N5-210 shot down by anti-aircraft fire over Langgoer.

24 August: 18 Squadron's N5-169 shot down by anti-aircraft fire during offensive shipping search near Tanimbar Island. Aircraft searching for missing Mitchell strafed two enemy vessels in Hom Bay.

1 September: co-ordinated night attack on Langgoer, 18 Squadron's N5-214 shot down by anti-aircraft fire.

2 September: 2 Squadron's A47-12 ditched near Perron Island due to fuel shortage. Mitchell A47-6 crashed in locality while searching for crew of A47-12.

12 September: another strike on Langgoer.

B-25J A47-46, one of the last Mitchells delivered to the RAAF in mid 1945. It did not see service with No 2 Squadron and was scrapped in 1950. (via Neil Mackenzie)

15 September: further strike on Langgoer with 18 Squadron's N5-222 shot down by anti-aircraft fire. This was 18 Squadron's last combat loss of the war.

No 2 Squadron lost a further six Mitchells during the course of 1944, A47-3, -8 and -33 went missing on operations, A47-2 crashed on takeoff from Hughes, A47-9 crashed on landing at the same strip and A47-11 made a forced landing on Bathurst Island after being damaged by anti-aircraft fire and was converted to components.

Non-operational accidents claimed two more of the RAAF's Mitchells during the course of 1944: A47-20 became the first loss in June 1944 when its brakes failed at a Repair Depot and the damage sustained resulted in the aircraft being reduced to spares, and in December A47-20 crashed during bombing practice at the Torquay Firing Range. The aircraft was not on 2 Squadron's strength at the time.

An unusual activity in which No 2 Squadron indulged during the second half of 1944 was the dropping of supplies to intelligence officers operating on Japanese occupied Timor. These operations were of vital importance and required patience and superb timing as more often than not the drop zone was very close to Japanese positions. The supplies were dropped in 'storpedo' containers.

Anti Shipping Specialists

The two Mitchell squadrons of 79 Wing enjoyed considerable success in the anti shipping role, between them sinking or badly damaging 22 vessels in October 1944, 32 in November (including four small freighters) and a further 29 in December plus many more at other times.

Some idea of this activity and results achieved are illustrated by this extract from the RAAF Official History which discusses a series of sweeps in the Sumba/Sawu/Roti Islands area (off Timor) between December 15 and 20, 1944: "Four Mitchells from No 18 Squadron attacked a 400 ton vessel with four motor vehicles on its deck near Waingapu, forcing it into shallow water. Later, four Mitchells from No 2 Squadron destroyed the vessel by mast-height bombing and strafing.

"On 16 December, a 70 ton lugger was destroyed by the Dutch pilots off Sawu. On the 18th, near Roti Island, Flying Officer (W G) Coward and his crew, of No 2 Squadron, scored a direct hit on a 150 ton vessel which blew up.

"On the 20th the Dutch pilots sank a 900 ton vessel, the *Shoeki Maru* off the Schildpad Islands in the Banda Sea [north of Timor]. A small convoy, travelling at 8 knots and comprising two motor vessels with an escorting torpedo boat had been sighted by reconnaissance aircraft. No 18 Squadron was ordered to attack it and they sent four Mitchells. Three of these found the convoy and shadowed it. The Torpedo boat tried to get ahead of the larger vessel (*Shoeki Maru*) to cover it, while the heavily armed smaller vessel remained at the rear.

"The three Mitchells, diving from 1,200 feet to mast height, attacked and scored direct hits. The bridge disappeared and the vessel caught fire. The Mitchells left it on fire from stem to stern and with the stern rising in the water. Many Japanese jumped into the water to escape the flames before the vessel sank".

The two Mitchell squadrons combined for a co-ordinated attack on a particular vessel Waingapu, Sumba Island on 6 November. Involved were four aircraft from No 2 Squadron and 12 from No 18 and the ordering of the raid resulted from the reporting of two motor vessels in the harbour by a reconnaissance aircraft.

The four Australian Mitchells, led by Flt Lt P J Norriss, were selected to attack the targets in the harbour from low level, while the Dutch aircraft (Major Boot leading) would attack Waingapu town and anti-aircraft positions from medium level so as to divert their attention from the four Australian Mitchells.

A 300 ton vessel was discovered in the harbour and it was regarded as being heavily damaged by its attackers. It wasn't until after the war that Japanese records confirmed that Special Submarine Chaser No 118 had indeed been sunk as a result of the raid. As was usually the case, the Mitchells delivered a well planned and accurate attack.

A Quiet Conclusion

By the beginning of 1945 the Japanese were steadily withdrawing from the areas which the Mitchells had spent most of their war, but there were still plenty of shipping targets to go after. A contribution to a major prize was claimed on 6 April 1945, the result of a joint effort by the Mitchells of Nos 18 and 2 Squadrons, and Liberators from Nos 21 and 24 Squadrons RAAF.

The prize was the Japanese light cruiser *Isuzu*, which had been spotted as part of a small convoy steaming in a south-easterly direction in the Flores Sea, heading for Koepang on Timor where it was assumed a short stop would be made to pick up evacuees. The convoy was photographed by Mosquitos of No 87 Photographic Reconnaissance Flight and then shadowed by Catalinas from No 43 Squadron before being attacked shortly after leaving Koepang in the early hours of the morning two days later.

Operating at the limit of their range, 18 Squadron's Mitchells attacked first, scoring two direct hits on the cruiser with 500lb bombs. No 2 Squadron failed to hit the *Isuzu* as did the following Liberators, which had to withstand stout resistance from anti-aircraft fire and enemy fighters. Two Liberators were shot down and although the damaged *Isuzu* escaped, it was finished off by an American submarine the following day.

A 79 Wing Mitchell taxies out through the heat and dust of its Northern Territory base. (via Neil Mackenzie)

Two shots of the TB-25J Mitchell brought to Australia in December 1983 by Aero Heritage, a Melbourne based consortium. The Mitchell was purchased by the Australian War Memorial and flown until late 1987 when it was put into storage. The photo (above) shows the Mitchell in the US Navy colours in which it arrived in Australia and was taken at Canberra in January 1984. The lower photo illustrates its later markings as No 2 Squadron's A47-31/KO-P. Its Australian civil registration was the appropriate VH-XXV. (David Daw/Stewart Wilson)

The final months of the war were frustrating for the Dutch squadron, which was keen to move to a forward area of operations. After several months of rumour, the beginning of 1945 heralded a move to Jacquinot Bay in New Britain for all the components of No 79 Wing and preparations for the move of the squadrons began as early as February with personnel and equipment making their way there over the next three months.

Despite this and after a period of relatively quiet operations from northern Australia, the squadron never moved to New Britain in its entirety for in May that decision was cancelled and 79 Wing activities stopped altogether at the same time. In May it was decided to move the Wing to Morotai Island to join the 1st Tactical Air Force, but even this was rescinded (after advance elements had already moved there), causing considerable frustration and even a degree of resentment within Nos 18 and 2 Squadrons. Officially, there was insufficient room and facilities at busy Morotai to accommodate the Mitchell squadrons.

Finally, in July, 79 Wing was assigned to Balikpapan to support the Borneo campaign but by the time No 18 Squadron's aircraft arrived there on 14 August 1945, the capitulation of the Japanese was only one day away. No 2 Squadron was ready to go fours days later. Neither squadron therefore had the opportunity to take part in the Borneo campaign and in fact had seen no action for some three months.

The end of hostilities saw the two Mitchell squadrons carrying out search, leaflet dropping, supply dropping to prisoners of war, special reconnaissance and prisoner of war evacuation flights. During this period 18 Squadron suffered its final operational loss of the war when N5-255 crashed at Malino, Macassar on 4 September. All aboard were killed and the crash was the 21st fatal operational loss suffered by the squadron since its formation at Canberra in 1942. The cost in human terms had been 112 Dutch and Australian airmen.

No 2 Squadron lost two aircraft in the August/September 1945 period; A47-37 on 14 August which went missing on a travel flight to Biak and A47-19, which crashed whilst escorting a Catalina from Balikpapan on 16 September. The final RAAF Mitchell casualties were recorded in October when 2 Squadron's A47-23 was damaged during servicing and scrapped, while B-25J A47-40 (which had not seen squadron service) crash landed after takeoff from Laverton in Victoria.

On 8 September, the Mitchells of Nos 18 and 2 Squadrons provided aerial cover to the surrender of Japanese commanders on board HMAS *Burdekin*, a *Rother* class frigate, their final act of World War II.

The war was now formally over, but 18 Squadron's activities continued as the Netherlands East Indies was gradually reoccupied by first the British and then the Dutch. On 15 January 1946 No 18 Squadron passed from RAAF to Dutch control, ending an important association formed from necessity in a time of war. It had been a successful partnership which had written a new chapter of Australian military aviation history.

18 Squadron didn't rest after the war against Japan had been won. The Indonesian uprising of 1946 kept it busy and, still flying Mitchells, it continued fighting for the next three years until an Armistice

One other Mitchell was in Australia in 1992, the Darwin Aviation Museum's ex USAAF B-25D 41-30222 "Hawg-mouth". The aircraft's restoration is well underway, with the forward fuselage/wing centre section area externally complete by late 1991. This Mitchell has a strong connection with northern Australia, serving as it did with the 380th Bombardment Group in the area and force landing 75 miles (120km) west of Tennant Creek NT in January 1945. It was recovered by the Darwin Aviation Historical Society in 1974. (Stewart Wilson)

Another shot of the Australian War Memorial's Mitchell before it was grounded, its wings removed and placed into what will undoubtedly be long term storage.

was signed in July 1949. The Republic of Indonesia was declared in December 1949 and No 18 Squadron finally disbanded in June 1950.

No 2 Squadron ceased operations on 14 November 1945 while still based at Balikpapan and a move back to Australia and disbandment was not too far away. By the end of December the squadron had no aircraft and a personnel strength of just two officers and three airmen and they were stationed at Laverton, Victoria, where the squadron had first been formed over nine years previously. In May 1946 No 2 Squadron RAAF was – for the moment – disbanded, although its surviving aircraft for the most part lingered in storage until March 1950 when they were sold through disposals for scrapping.

RAAF NORTH AMERICAN B-25 MITCHELLS

Abbreviations: soc – struck off charge; t/o – takeoff; NEIAF – Netherlands East Indies Air Force; USAAF – United States Army Air Force; ops – operations; cvtd – converted; AA – anti aircraft.

Notes: The RAAF's first 39 Mitchells (A47-1 to 39) were received ex-Netherlands East Indies Air Force, having previously served with or having been allocated to the NEIAF's No 18 Squadron. The 'NEIAF/USAAF No' column reflects the aircraft's previous identity or identities before being issued to the RAAF.

RAAF No	Model	NEIAF/USAAF No	Deliv	Disposal/Remarks
A47-1	B-25D	N5-183/42-87607	04/44	soc 03/50
A47-2	B-25D	N5-187/43-3422	04/44	crashed on landing Hughes Strip NT 12/44
A47-3	B-25D	N5-181/43-3423	04/44	lost on ops north coast Timor 09/44
A47-4	B-25D	N5-189/43-3424	04/44	converted to components 12/46
A47-5	B-25D	N5-192/43-3426	04/44	soc 03/50
A47-6	B-25D	N5-193/43-3427	04/44	crashed 09/44 whilst searching for crew of A47-12 near Perron Is
A47-7	B-25D	N5-194/43-3607	04/44	soc 03/50
A47-8	B-25D	N5-195/43-3613	04/44	lost on ops 11/44
A47-9	B-25D	N5-196/43-3621	04/44	crashed and burnt on t/o Hughes Strip NT 12/44
A47-10	B-25D	N5-197/43-3623	04/44	soc 03/50
A47-11	B-25D	N5-198/43-3624	04/44	crash landed Bathurst Is 12/44 after AA damage, cvtd to components
A47-12	B-25D	N5-199/43-3625	04/44	ditched near Perron Is 09/44, out of fuel
A47-13	B-25D	N5-200/43-3626	04/44	crashed during bombing practice 08/44
A47-14	B-25D	N5-201/43-3766	04/44	soc 03/50
A47-15	B-25D	N5-202/43-3767	04/44	soc 03/50
A47-16	B-25D	N5-203/43-3768	04/44	converted to components 08/45
A47-17	B-25D	N5-204/43-3769	04/44	soc 03/50
A47-18	B-25D	N5-205/43-3770	04/44	soc 03/50

RAAF No	Model	NEIAF/USAAF No	Deliv	Disposal/Remarks
A47-19	B-25D	N5-206/43-3790	04/44	crashed 09/45 whilst escorting Catalina from Balikpapan
A47-20	B-25D	N5-207/43-3791	04/44	damaged at Repair Depot 06/44 when brakes failed, cvtd to components
A47-21	B-25D	N5-213/43-3789	06/44	soc 03/50
A47-22	B-25D	N5-190/43-3830	06/44	soc 03/50
A47-23	B-25D	N5-212/43-3832	06/44	damaged during servicing 10/45, cvtd to components
A47-24	B-25D	N5-216/43-3867	06/44	crashed Torquay Firing Range 12/44
A47-25	B-25D	N5-215/43-3869	06/44	soc 03/50
A47-26	B-25J	N5-220/43-27689	06/44	soc 03/50
A47-27	B-25J	N5-219/43-27691	06/44	soc 03/50
A47-28	B-25J	N5-224/43-27927	07/44	soc 03/50
A47-29	B-25J	N5-225/43-27928	07/44	soc 03/50
A47-30	B-25J	N5-229/43-28185	07/44	soc 03/50
A47-31	B-25J	N5-231/43-28183	07/44	soc 03/50
A47-32	B-25J	N5-227/43-28181	08/44	soc 03/50
A47-33	B-25D	N5-175/42-87259	08/44	lost on ops 12/44
A47-34	B-25D	N5-186/42-87608	08/44	soc 03/50
A47-35	B-25D	N5-168/42-87416	08/44	soc 03/50
A47-36	B-25D	N5-171/42-87255	08/44	soc 03/50
A47-37	B-25D	N5-174/42-87258	08/44	lost en route Biak 08/45
A47-38	B-25J	N5-232/44-29021	09/44	soc 03/50
A47-39	B-25J	N5-235/44-29024	09/44	soc 03/50
A47-40	B-25J	– /44-30888	04/45	crashed after t/o Laverton Vic 10/45
A47-41	B-25J	– /44-30889	04/45	soc 03/50
A47-42	B-25J	– /44-30890	04/45	soc 03/50
A47-43	B-25J	– /44-30895	04/45	soc 03/50
A47-44	B-25J	– /44-30896	04/45	soc 03/50
A47-45	B-25J	– /44-30897	05/45	soc 03/50
A47-46	B-25J	– /44-31255	05/45	soc 03/50
A47-47	B-25J	– /44-31254	05/45	soc 03/50
A47-48	B-25J	– /44-31253	06/45	soc 03/50
A47-49	B-25J	– /44-86859	08/45	soc 03/50
A47-50	B-25J	– /44-86855	08/45	soc 03/50

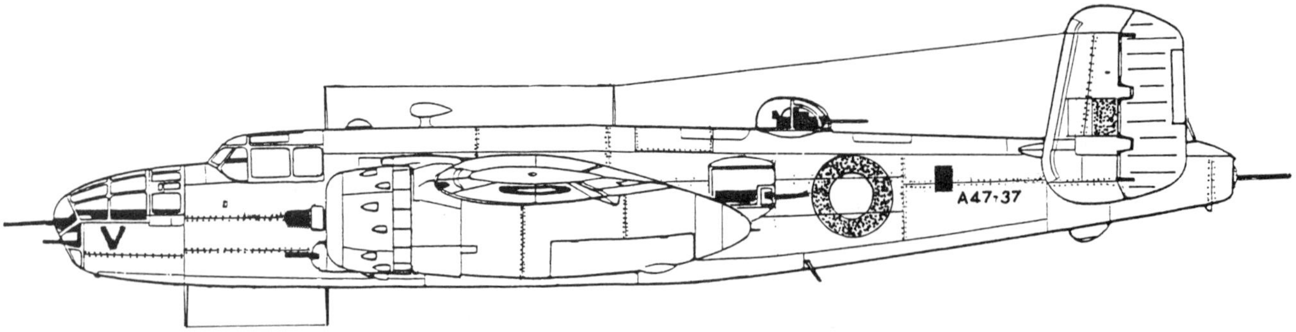

B-25D A47-37/V of No 22 Sqn
Natural metal overall. Black code (limited to the V on the nose) and serial. Olive drab propeller hub.

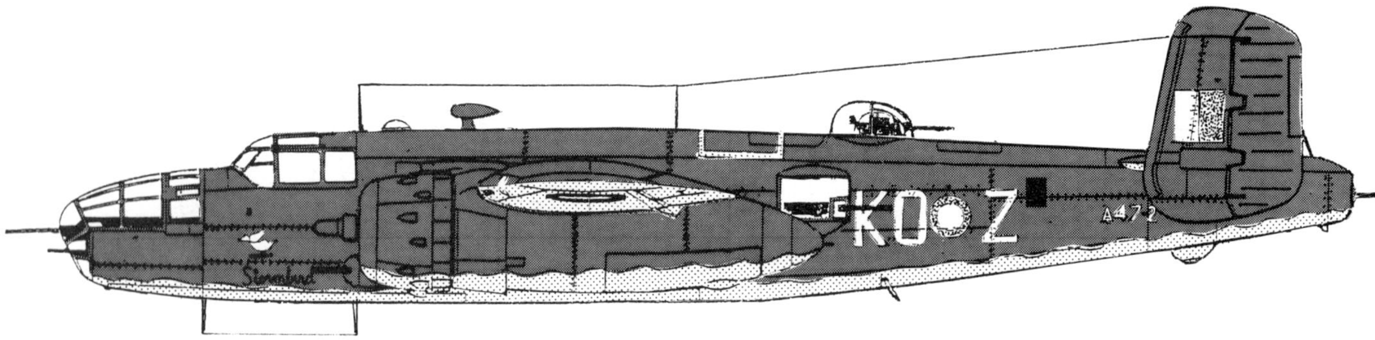

B-25D A47-2/KO-Z of No 22 Sqn
Olive drab upper surface with neutral grey undersurface. Light grey code and serial. Off-white tape around escape hatch on upper fuselage. Artwork below cockpit consists of a white bird with light brown wings carrying a black bomb. Name "Stormbird" below it is yellow.

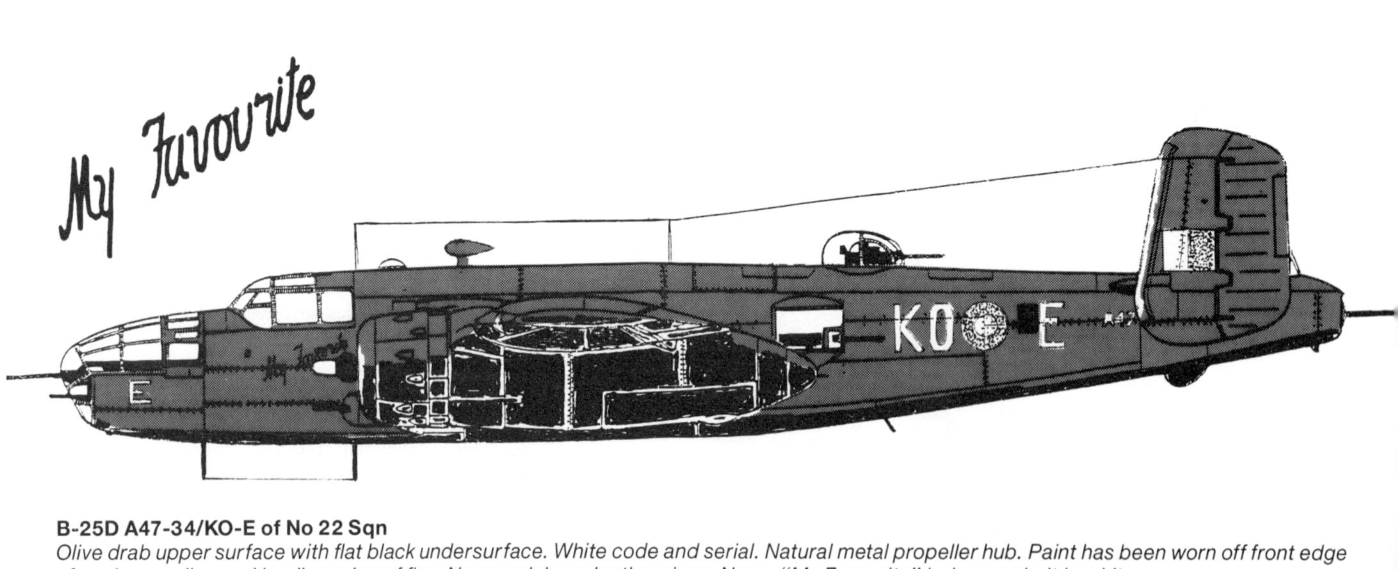

B-25D A47-34/KO-E of No 22 Sqn
Olive drab upper surface with flat black undersurface. White code and serial. Natural metal propeller hub. Paint has been worn off front edge of engine cowling and leading edge of fins. No roundels under the wings. Name "My Favourite" below cockpit in white.

CONSOLIDATED
B-24 LIBERATOR

Liberator A72-116, one of 83 B-24Ls supplied to the RAAF. This aircraft belongs to No 7 Operational Training Unit at Tocumwal NSW. (RAAF)

Of the many interesting aspects of studying aviation history, it's sometimes intriguing to look at different types of the same era and role and to examine which of them became the best known aircraft, the 'household name'. In the case of American heavy bombers of World War II, most will think of the Boeing B-17 Fortress before its colleague, the Consolidated B-24 Liberator. The B-17 has always attracted the 'glamour' and the publicity; it has been immortalised in books and on film – it was *the* American bomber of the war.

Yet the B-24 Liberator perhaps should have attracted the lion's share of attention. It was a later and more innovative design than the B-17, it was manufactured in considerably larger numbers, it was much more versatile (with bomber, maritime patrol/anti submarine, passenger, freighter, tanker, photographic reconnaissance and other variants) and it served on all of World War II's battle fronts in the colours of 15 Allied nations.

For Australia, the Liberator served the RAAF in large numbers for a heavy bomber, the first of 287 which were taken on charge entering service in early 1944 and leaving it just four years later. During that time, Liberators equipped seven RAAF operational squadrons and played a major role in the latter stages of the war against Japan, particularly in the Borneo campaign. The Liberator was the only heavy bomber to be taken on charge in operational numbers by the RAAF during World War II, bearing in mind that the 'Australian' bomber squadrons operating in Europe flew RAF aircraft. At one stage, the production of the Liberator in Australia under licence was even considered.

Towards the Liberator

The Munich crisis of 1938 alerted many nations to the suddenly very real possibility of a major war. Even the isolationist and neutral United States was prompted into some action regarding the state of its armed forces including the Army Air Corps. The immediate pre-war period saw the start of several combat aircraft designs which would go on to fight between 1939 and 1945 and form the basis of the huge industrial effort which resulted in the manufacture of hundreds of thousands of aircraft in that period.

The Consolidated Model 32 four engined heavy bomber was one of those, its final development stemming from this period although design studies for such an aircraft had been undertaken by Consolidated as early as 1935. By the late 1930s the benchmark was the Boeing B-17C Fortress, and the aim was to produce a bomber capable of carrying the B-17's load further, higher and faster.

It wasn't until January 1939 that the Model 32 began moving towards becoming a firm programme. In that month, Consolidated was invited to prepare a design study for a bomber with performance superior to that of the Fortress. The specification was demanding for its time, calling for a maximum speed in excess of 300mph (480km/h), a range of 3,000 miles (4,800km) and a service ceiling of 35,000 feet (10,668m).

Under Consolidated's design chief engineer Isaac 'Mac' Laddon, the key to the Model 32's design was its quest for range, and the means to achieve that came via its wing design, the high aspect ratio, high lift 'Davis' wing.

This for its time radical wing design fell into Consolidated's lap in 1937 with the appearance on its doorstep of one David R Davis, engineer and inventor. Davis spoke to Consolidated's co-founder, well known designer Reuben Fleet, and presented him with some radical ideas about aerofoils which, it was claimed, would offer 25 per cent less profile drag at high speeds and 10 per cent less at low speeds than conventional aerofoils. Davis' aerofoil featured a reflex curve on its undersection aft of where the rear spar would be located. This would enable heavy box spars to be used along with a high aspect ratio and wing loading, resulting in a stiff structure and plenty of space for fuel.

Mac Laddon tested the aerofoil in a wind tunnel and discovered its efficiency was being measured at 102 per cent, two per cent more than was theoretically possible! At first it was thought the measuring of data was faulty, but even allowing for some errors there, the Davis aerofoil was obviously very efficient.

It was decided, therefore, to build a patrol flying boat around the wing, bearing in mind that by 1938 Consolidated's PBY Catalina boat was only a year away from its planned end of production and a successor was considered desirable. World events substantially extended the Catalina's production run!

The resulting Model 31 flying boat was a one-off aeroplane which first flew in May 1939. A very large aircraft for a twin, it featured a deep and bulky fuselage and its Davis wing incorporated Fowler area increasing flaps. Testing of the Model 31 revealed that the 110 feet (33.53m) span Davis wing worked well in practice as well as in theory and it was therefore used as the basis of the forthcoming Model 32 bomber design with little change. Several other of the 31's features would also appear on the Model 32/Liberator including the Fowler flaps (which promised better airfield performance), the canopy and the horizontal tail surfaces with their endplate twin fins and rudders.

The prototype Consolidated Model 32 (USAAC XB-24) at its roll out in December 1939. The aircraft carried the USAAC serial number 39-556.

The XB-24 recorded its maiden flight on 29 December 1939 from Lindbergh Field, San Diego.

Overhead view of the XB-24 emphasises the efficient high aspect ratio Davis wing. Wind tunnel testing of the wing section revealed a theoretically impossible efficiency rating of 102 per cent!

The XB-24

Consolidated's submissions to the USAAC for the Model 32 were met favourably, even though the design could have been regarded as radical in many ways with its tricycle undercarriage, shoulder mounted Davis wing of the 'wet' type (in which entire wing sections were sealed to form fuel tanks rather than the more conventional separate fuel cells within the structure) and a very deep fuselage incorporating a central bomb bay which could carry a 4,000lb (1,814kg) load in each of its two separate compartments. The compartments were connected by a central catwalk which also served as the fuselage keel beam and the bomb bay doors comprised roller shutter style structures which retracted upwards into the fuselage, thus creating considerably less drag than the usual outwards opening doors.

The shoulder wing design of the Model 32 necessitated long main landing gear legs which retracted outwards into the wing undersides, while the short steerable nosewheel retracted forwards into the fuselage. All undercarriage units were fitted with single wheels.

Structurally, the all metal Model 32 was fairly conventional with a cantilever two spar wing built in three sections (centre and two outer sections with detachable tips), aluminium alloy semi monocoque fuselage and cantilever tail unit. The ailerons were metal framed with fabric covering, as were the elevators while the rudders had metal leading edge covering and fabric behind that. The remainder of the airframe was covered in light alloy skinning.

Power would be provided by four 1,200hp Pratt & Whitney R-1830 Twin Wasp 14 cylinder two-row radial engines in several versions throughout the Liberator's production life, driving Hamilton Standard Hydromatic constant speed, fully feathering three bladed propellers of 11ft 7in (3.53m) diameter. As initially proposed, defensive armament would comprise four single hand operated 0.50in machine guns mounted in the nose, two waist positions and the tail, the latter being a new feature for an American aircraft of the time. As the Liberator developed, the number of guns would be doubled and powered turrets would be fitted in the nose, dorsal, tail and ventral positions.

Another figure which would increase was the number of crew. The original Model 32 design featured accommodation for seven but by later versions carried a crew of ten or eleven comprising two pilots, navigator, bombardier, flight engineer/radio operator and four or five gunners.

A contract for a single Model 32 prototype was placed on 30 March 1939 under the USAAC designation XB-24 and included a clause which required the new bomber to make its maiden flight within nine months of the contract being signed. Manufacture immediately got underway at Consolidated's San Diego plant and other orders quickly followed. Within a month of the original contract being placed the USAAC ordered a further seven YB-24 service evaluation aircraft while the rapidly worsening situation in Europe during the course of 1939 prompted France to order 120 examples of the new bomber sight unseen. The British added 164 aircraft to the orderbook early in 1940 and as was the case with several American combat aircraft of the period, it was these orders which got the project going and saw firm production orders placed before the US military had done so.

The French never received their Liberators due to the German invasion but the British did and became the first operator of the type. An interesting sidelight is why the name "Liberator" was chosen for the new bomber. According to Reuben Fleet (while explaining the name to the British Air Ministry): "We chose the name Liberator because this airplane can carry destruction to the heart of the Hun, and thus help you and us to liberate those nations temporarily finding themselves under Hitler's yoke". Fine sentiments and not untypical in the USA at the time, despite that country's official neutrality.

The first Liberator variant to enter US Army service was the B-24A, just nine of which were received in 1941. Like their British counterparts of the time, the B-24As were used as transports.

With Consolidated's chief test pilot Bill Wheatley at the controls, the XB-24 flew for the first time on 29 December 1939 from San Diego's Lindbergh Field. The date, incidentally, was just one day short of the limit specified in the USAAC contract. In its original form, the XB-24 was powered by four 1,200hp R-1830-33 engines with mechanical two speed superchargers fitted. Gross weight was initially 38,300lb (17,373kg) and testing revealed a maximum speed of 273mph (439km/h), perhaps disappointingly some 50mph (80km/h) slower than the contemporary B-17C Fortress which was fitted with turbo-supercharged engines.

On the other hand, the XB-24 exhibited better payload-range characteristics than its Boeing rival from the word go: with full bomb load it could fly 200 miles (320km) further than the B-17; with its forward bomb bay fitted with auxiliary fuel tanks and its rear bay filled with a useful bomb load, it had a range 600 miles (965km) greater than a similarly equipped B-17 which could carry no load at all.

The XB-24 was subject to considerable modification during its life including the fitting of R-1830-41 engines with General Electric B-2 turbo-superchargers for considerably enhanced performance at altitude, increased tailplane span and the replacement of the wet wing fuel system with individual self-sealing fuel cells. The modifications caused the first of many increases in the Liberator's maximum weight, in this case to over 41,000lb (18,600kg).

Redesignated the XB-24B, the aircraft flew for the first time in its new form on 1 February 1941.

Into Production

While testing of the XB-24/B continued, production of the bomber was meanwhile getting underway at San Diego, beginning with the seven YB-24 service test aircraft during 1940. These were similar in specification to the XB-24 in its original form, differing only in detail. With the collapse of France, aircraft originally intended for that country were allocated to the Royal Air Force, and it was for Britain that most of the early newly named Liberators were built.

Six of the YB-24 evaluation aircraft were the first to go to the RAF under the company designation LB-30A, the 'LB' prefix standing for *L*iberator built to

First Liberator version to enter service was the LB-30A, six of which were delivered to the Royal Air Force from late 1940. Found to be unsuitable for combat over Europe, they were successfully employed as unarmed transports on the Trans-Atlantic Ferry Return Service.

Royal Air Force Coastal Command was a major user of the Liberator in several versions, the aircraft being the first capable of closing the 'Atlantic Gap' in which German submarines could roam unhindered. Illustrated is a Liberator III (top) and Liberator GR.V (bottom), the latter displaying its retractable Air-to-Surface-Vessel (ASV) radar. Both were based on the B-24D. (via Neil Mackenzie)

British specifications. The aircraft were delivered to the RAF in December 1940 but were considered to be unsuitable for European combat conditions due to their lack of self-sealing fuel tanks. Their long range and relatively capacious fuselages were an asset, however, and from March 1941 the Liberators were successfully employed as unarmed transports on the Trans-Atlantic Ferry Return Service. This pioneering transport service contributed much to postwar long range services and at the time the LB-30 was the only aircraft capable of tackling the 2,994 mile (4,818km) Prestwick (Scotland) to Montreal (Canada) route non-stop, cruising at 193mph (310km/h).

Twenty LB-30B Liberator Is followed on the production line and were delivered to the RAF in the first half of 1941. These were originally intended to go to the United States Army Air Corps (renamed US Army Air Force in 1940) as B-24As but were allotted to the RAF when war broke out. They were fitted with the then standard RAF bomber defensive armament of six 0.30in Browning machine guns, with one in the nose, two in the tail, one in each waist position and one in the ventral position.

The LB-30B/Liberator I was the first variant of Consolidated's bomber to see active service, joining the RAF Coastal Command's No 120 Squadron for maritime patrol and anti submarine duties where the aircraft's long range and useful offensive load made it ideal. The LB-30Bs were the first aircraft with the ability to close the 'Atlantic Gap', where

The RAF's Liberator II had no USAAF equivalent but introduced some new features including self sealing fuel tanks and a lengthened nose. This is the first of them, AL503, which crashed before delivery to the RAF. (Bob Livingstone)

German U-Boats had free reign and caused heavy losses to the vital convoys travelling from North America to Britain.

In Coastal Command service the LB-30Bs were extensively modified from the configuration in which they were delivered, changes including the incorporation of four forward firing 20mm cannon mounted in blisters under the nose and the fitting of ASV radar and the myriad of aerials associated with that equipment.

The USAAF received just nine B-24A Liberators in 1941, these differing from the British LB-30Bs mainly in the fitting of 0.50in machine guns in some positions, a different sub-variant of the R-1830 Twin Wasp engine, and a higher maximum takeoff weight. Like the original six LB-30As delivered to Britain, the USAAF's B-24As were put to work as transports and made several notable long distance flights carrying President Roosevelt's senior cabinet members on diplomatic missions around the world.

The Liberator II for the Royal Air Force had no USAAF equivalent and introduced several "firsts" to the Liberator programme, including the lengthened (by 2ft 7in) nose which would become standard on subsequent Liberators, powered turrets and self-sealing fuel tanks. Deliveries to the RAF began in August 1941 after a two month delay resulting from the investigation into a pre-delivery accident at San Diego which destroyed the first example and killed all on board including Consolidated's chief test pilot William Wheatley. It was found that a loose bolt had jammed the elevators of the Liberator.

The RAF took delivery of 139 Liberator IIs which served with three Coastal and two Bomber Command squadrons, the latter in the Middle East. Some of these Liberators were delivered as unarmed transports under the designation LB-30. Perhaps the best known of these was AL504 "Commando", which became the personal aircraft of British Prime Minister WInston Churchill and was used by him on numerous journeys during the war.

The Liberator II was perhaps the first fully operational combat variant of the bomber. Its defensive armament comprised two Boulton Paul power operated turrets (in the tail and mid upper positions) each housing four 0.303in Browning machine guns and three further manually operated 303s, in the waist positions and nose. Curtiss Electric three bladed propellers replaced the previous Hamilton Standard unit, the maximum weight was 46,250lb (20,979kg) and maximum speed was 263mph (423km/h).

After the Japanese attack on Pearl Harbour in December 1941 and the USA's entry into the war, 75 Liberators IIs intended for the RAF were requisitioned for USAAF use and redesignated LB-30s. These were operated in both bomber and transport versions, the former replacing the Boulton Paul turrets of the British aircraft with Martin turrets each housing a pair of 50 calibre machine guns (single handheld 'fifties' occupying the other positions) and the latter featuring a solid nose, a faired over tailcone, the removal of all guns and a large cargo door on the port side of the rear fuselage.

Twelve of these LB-30s were deployed to Java in January 1942 to reinforce the 19th Bombardment Group (and thus became the first USAAF Liberators to see action) and the survivors of this group formed the earliest association between the Liberator and Australia, more of which later.

The Might of American Industry

The United States' entry into World War II sparked an amazing flexing of America's industrial muscle. Put on a war footing, US industry produced everything needed for the Allies' war effort in huge quantities and in a very short time. The Liberator was subject to perhaps the greatest such programme of all the aircraft when the B-24 Production Pool Programme was put into place in early 1942. This would see 18,482 (plus hundreds more as spares) Liberators manufactured up until May 1945 when the final example was handed over to the USAAF.

The programme involved five factories – Consolidated/San Diego (California), Consolidated/Fort Worth (Texas), Douglas/Tulsa (Oklahoma), Ford/Willow Run (Michigan) and North American/Dallas (Texas) – all of which came on line during 1942. All but the Douglas/Tulsa plant were used for both primary manufacture and assembly, while Douglas' facility was used only for assembly. To give some idea of effort involved, at its peak, Ford's Willow Run plant was producing 200 complete B-24s plus a further 150 in the form of 'knocked down' assemblies per month ... one aircraft every 100 minutes of the working day!

The production pool system worked well, except that it did cause problems with standardisation of components and equipment. With the same Liberator variant being produced in several different factories and different but similar variants emerging from these same plants, it could be a logistical nightmare for parts and repair depots around the world,

A B-24D Liberator of an RAF squadron in the India-Burma campaign displays some nose artwork undoubtedly influenced by Australian members of its crew. The B-24D was the first mass produced Liberator variant. (via Neil Mackenzie)

One of the best known Liberators was Mk.II/LB-30 AL504 'Commando', used by Winston Churchill as his personal transport. The photograph (top) shows the aircraft in its original guise while the British Prime Minister was visiting No 3 Squadron RAAF in the Western Desert; the lower shot shows the aircraft after modification to Liberator C.IX standards in 1944 with cabin windows and a single fin and rudder. (via Neil Mackenzie)

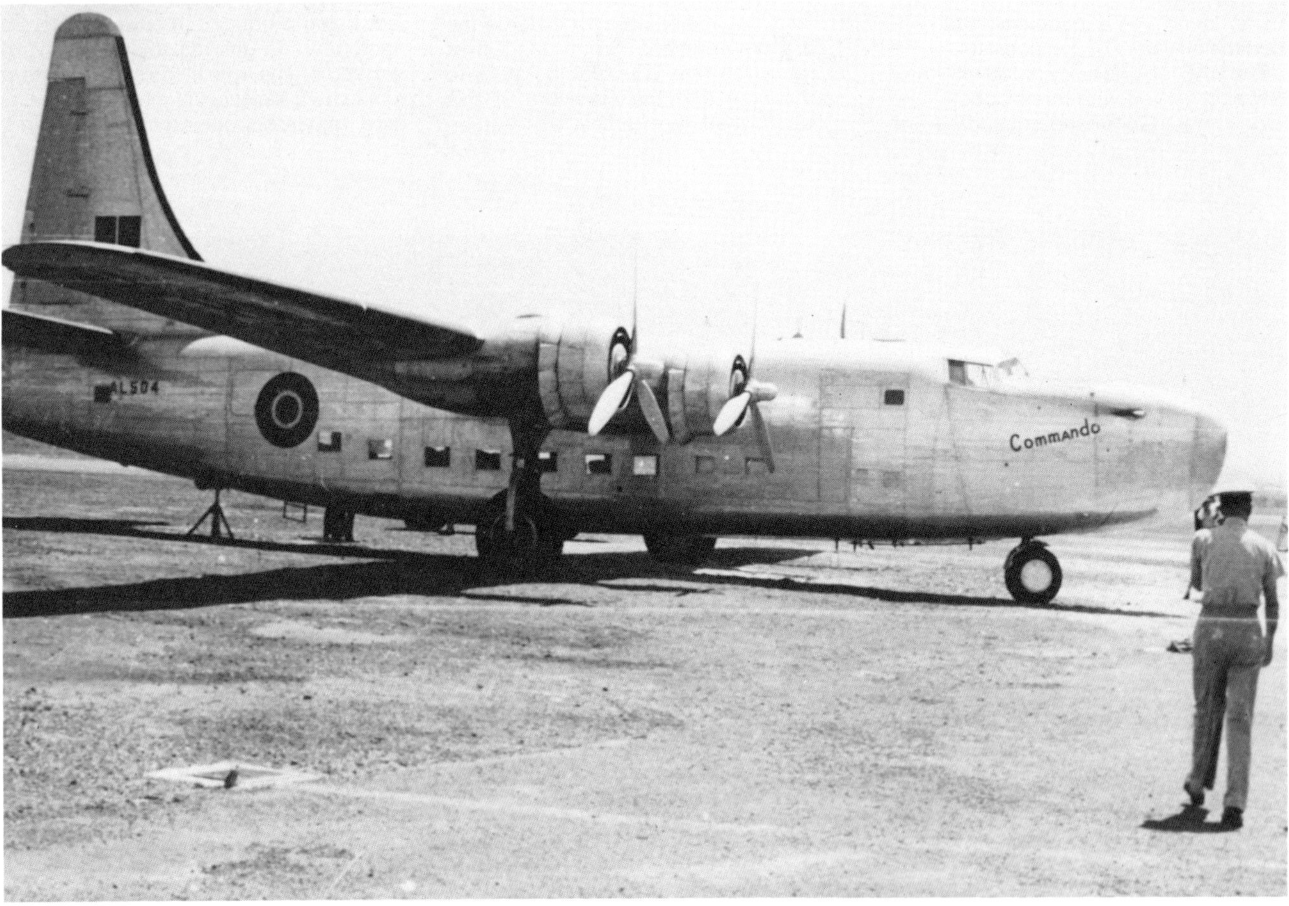

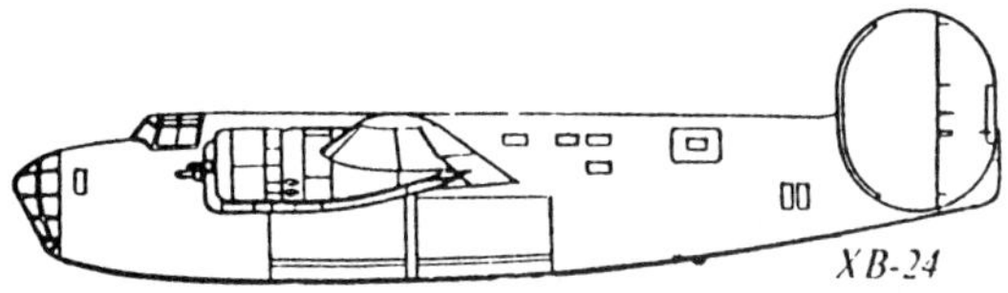

XB-24

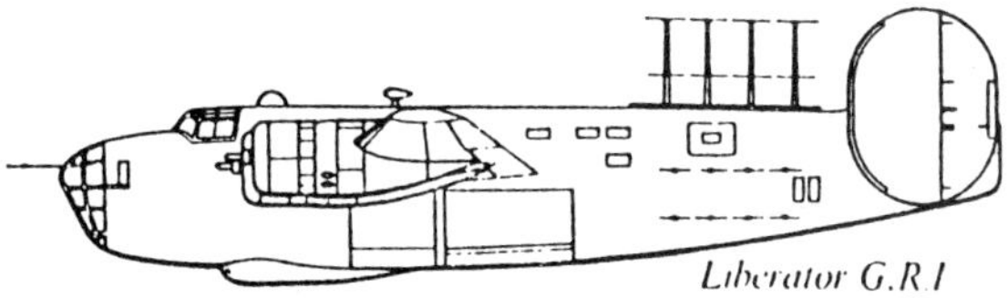

Liberator G.R.I

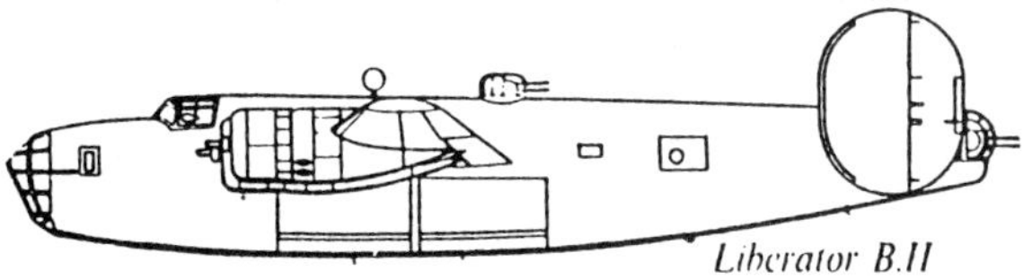

Liberator B.II

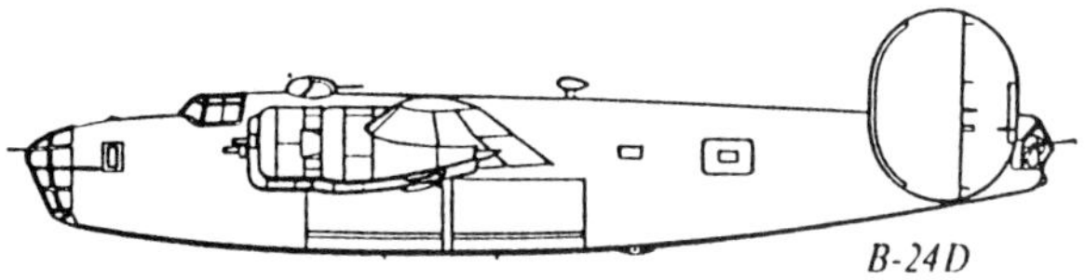

B-24D

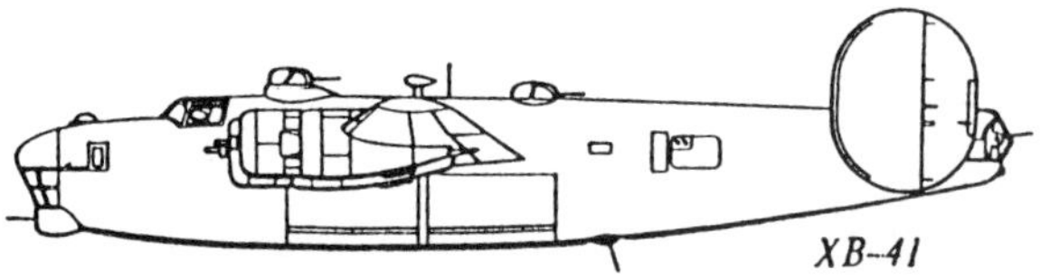

XB-41

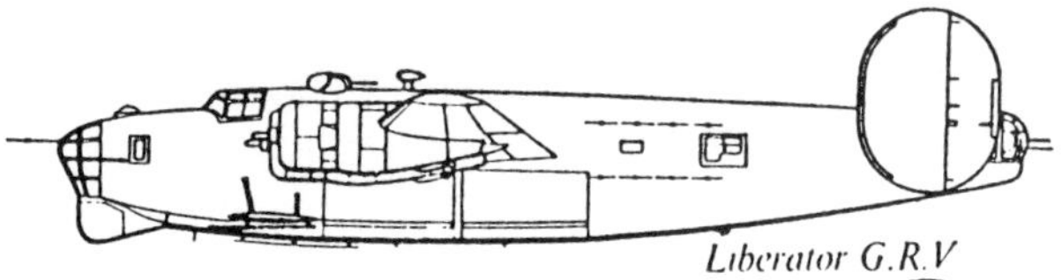

Liberator G.R.V

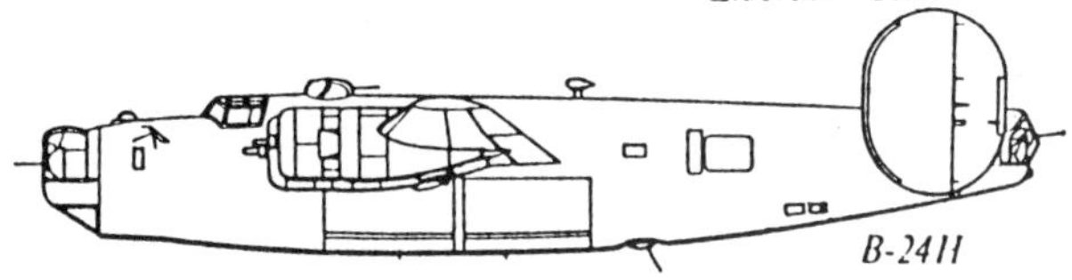

B-24H

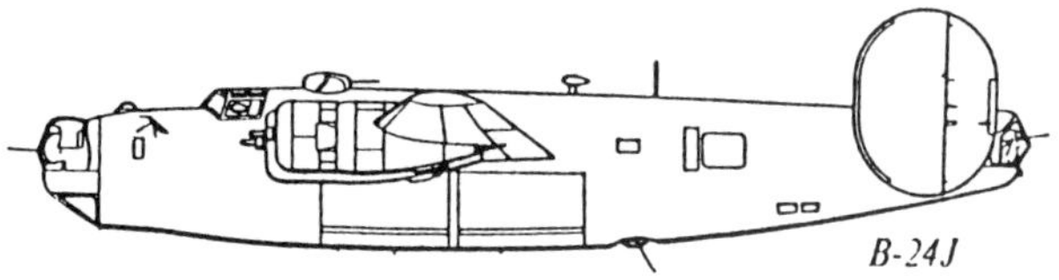

B-24J

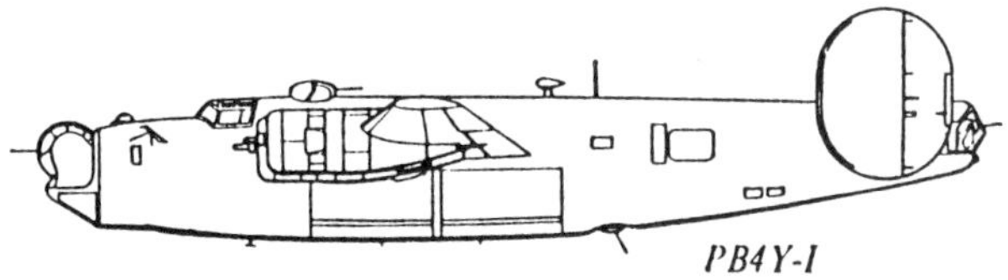

PB4Y-1

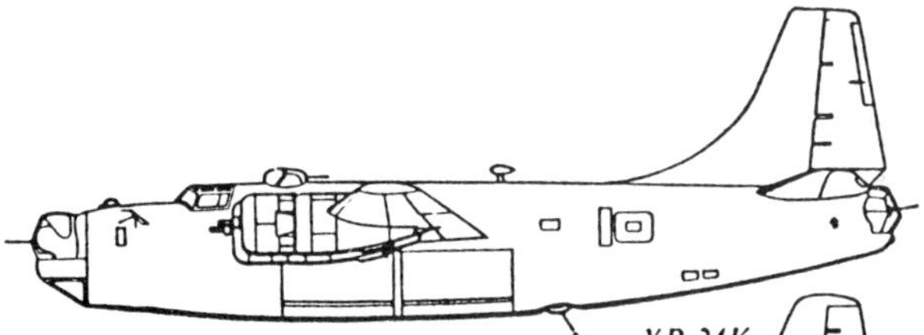

XB-24K

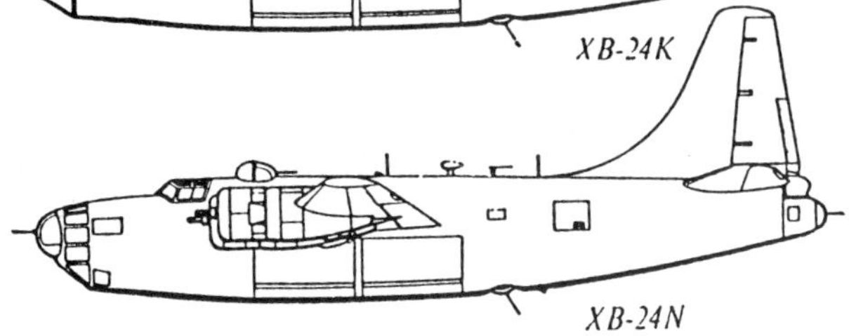

XB-24N

which had to carry spare parts covering the numerous possible permutations. There were in fact significant differences between the various production blocks of the same model Liberator and indeed some differences within a production block. Parts for aircraft built at different factories were often not interchangeable, and all four factories involved in primary manufacture would produce Liberators of similar variant but of often vastly different detail specification. Even the two Consolidated factories suffered from this.

But before mass production of the Liberator began, there was one more low volume variant, the B-24C. An interim model, only nine were delivered to the USAAF right at the end of 1941. Several significant features were introduced on the B-24C, some of which had previously been tested on the first Liberator in its XB-24B form. Among these were self-sealing fuel tanks and turbo-supercharged R-1830-41 engines, still rated at 1,200hp but capable of maintaining that to above 20,000 feet with the resultant increase in top speed to more than 300mph (483km/h).

The longer nose first introduced on the British Liberator II was incorporated and the engine cowlings took on the distinctive oval shape (when viewed from the front) which would be a feature of future Liberators. This shape resulted from the installation of air intakes on the side of the cowlings.

The B-24C reverted to Hamilton Standard propellers (as did all subsequent Liberators), and had a Martin upper turret housing two 0.50in machine guns, a similarly equipped Consolidated rear turret and single manual 'fifties' in the nose and waist positions.

The B-24D was the first massed produced Liberator, 2,738 of which were built from January 1942. Essentially similar to the B-24C, this Liberator was continually refined during its production life with variations in defensive armament prominent among them. The upper and tail turrets remained unchanged, as did the waist guns but the single ventral (or 'tunnel' gun) was replaced with a two-gun manned Sperry ball turret (similar to that fitted to the B-17E Fortress) after the 76th production aircraft. This was deleted after a time and the single tunnel gun reinstated.

The nose defensive armament initially comprised a single 0.50in machine gun but additional guns mounted in ball sockets in each side of the nose were later fitted. Normal maximum bomb load was 8,000lb (3,629kg) but with revisions to the bomb shackles design this could be increased to 12,800lb (5,806kg) over short ranges. This increased the maximum gross weight to 67,800lb (30,754kg) and later on there was a further increase in the maximum possible to 71,200lb (32,296kg). The normal takeoff weight for a B-24D carrying an 8,000lb bomb load was around 64,000lb (29,030kg).

The initial powerplant fitted was the turbo-supercharged R-1830-43 (all production Liberators from now would have turbo engines) and this was later replaced by the R-1830-65 which could maintain its 1,200hp rating up to 31,800 feet. Maximum speed was 303mph (487km/h) at 25,000 feet and at maximum cruise power the B-24D could fly a distance of 1,800 miles (2,896km) with normal maximum fuel of 1,978imp gal (8,992 l). Later B-24Ds could carry up to 3,009imp gal (13,679 l) of fuel if auxiliary outer wing and bomb bay tanks were fitted.

The B-24D was supplied to the Royal Air Force as the Liberator III (0.303in guns, Boulton Paul tail turret) and Liberator IIIA (American equipment). Some of these aircraft were heavily modified for use by the RAF's Coastal Command as the Liberator GR.V. They featured chin or retractable ventral radar radomes, a Leigh Light for the night detection of Submarines on the surface, increased fuel capacity (at the expense of armour and fuel tank protection) and in some cases, provision for eight rocket projectiles mounted on stub wings on the forward fuselage. Defensive armament typically comprised one 0.303in or 0.50in gun in the nose, twin 'fifties' in the Martin upper turret, four 303s in a Boulton Paul rear turret and twin 303s or single 50s in the waist positions.

Two Liberator variants which were very close to the specification of late

Two views of the world's only airworthy Liberator in 1992, the superbly restored B-24J 44-44052. Named 'All America', the Liberator carries the names or colours of every USAAF group which flew the type, or of individuals who flew it. (Jim Thorn)

production B-24Ds were the B-24E manufactured at Ford/Willow Run for final assembly there plus Consolidated/Fort Worth and Douglas/Tulsa (791 built from mid 1943) and the B-25G, just 25 of which were built at North American/ Dallas during 1943. Substantially greater numbers of an improved 'G' would shortly follow. Only one XB-24F Liberator was produced, this B-24D conversion featuring an experimental thermal airframe de-icing system.

Magnificent Multiplicity

A major weakness in the Liberator's defensive armoury was its vulnerability to frontal attack by enemy fighters, the original installation of a single manually operated machine gun in the nose offering next to no defence and the addition of two other similar nose guns helped little. A powered turret was required and several units began field modifications of their aircraft, installing the Consolidated rear turret in the noses of their B-24Ds. The modification was successful and these turrets were installed in many B-24Ds before they were sent to fight in the Pacific.

All the later production variants of the B-24 would feature nose turrets mounting a pair of 50 calibre machine guns of either Consolidated or Emerson design. Most Liberators from late 1943 featured four two-gun turrets – nose, ventral, mid-upper and tail – although there were variations on this. The waist guns were usually retained. Nose turrets were either Emerson or Consolidated, ventral turrets were Sperry, tail turrets Consolidated and upper turrets Martin. Some British Liberators of the basically similar B-24G/H/J series (RAF Liberator VI and VIII, US Navy PB4Y-1) had a Boulton Paul rear turret, meaning it was possible to have four different brands of turret on the same aircraft!

The B-24G/H/J series were really variations of the same aircraft, differing mainly in details of equipment due to their factory of origin. The B-24G (430 built including the 25 early examples mentioned above) was manufactured at North American/Dallas, the B-24H (3,100 built) manufactured at Ford/ Willow Run for assembly there as well as Fort Worth and Tulsa and the B-24J (6,678 built) was manufactured and assembled in all five factories. Deliveries of the B-24J began in August 1943, five months after Consolidated had merged with Vultee Aircraft to form Consolidated-Vultee, or Convair as it became known.

Apart from the installation of the nose turret, the B-24J and its close relatives featured (for the most part) R-1830-43 or -65 engines, and an improved autopilot, bomb sight and fuel transfer system. Later aircraft with -65 engines also had improved General Electric B-22 turbochargers which further enhanced performance at altitude. But there were other variations to add to the confusion including the incorporation of staggered waist gun positions and handheld but hydraulically assisted twin 'fifties' in the tail of some aircraft. This installation was known as the 'Stinger' and offered a greater field of fire to the tail gunner as well as saving about 200lb (91kg) in weight.

The weight of the Liberator increased with its development, to a normal maximum of 65,000lb (29,484kg) or overload of no less than 71,200lb (32,296kg). Performance highlights of the B-24J included a maximum speed of 300mph (483km/h) at 30,000 feet and a range of 1,700 miles (2,736km) cruising at 232mph (373km/h) while carrying a 5,000lb (2,268kg) bomb load.

Some front end detail of a bogged and probably embarrassed RAAF B-24M Liberator, A72-178. Note the oval shaped engine cowlings, a feature of all Liberators from the B-24C on, with turbosupercharged engines. (RAAF Historical)

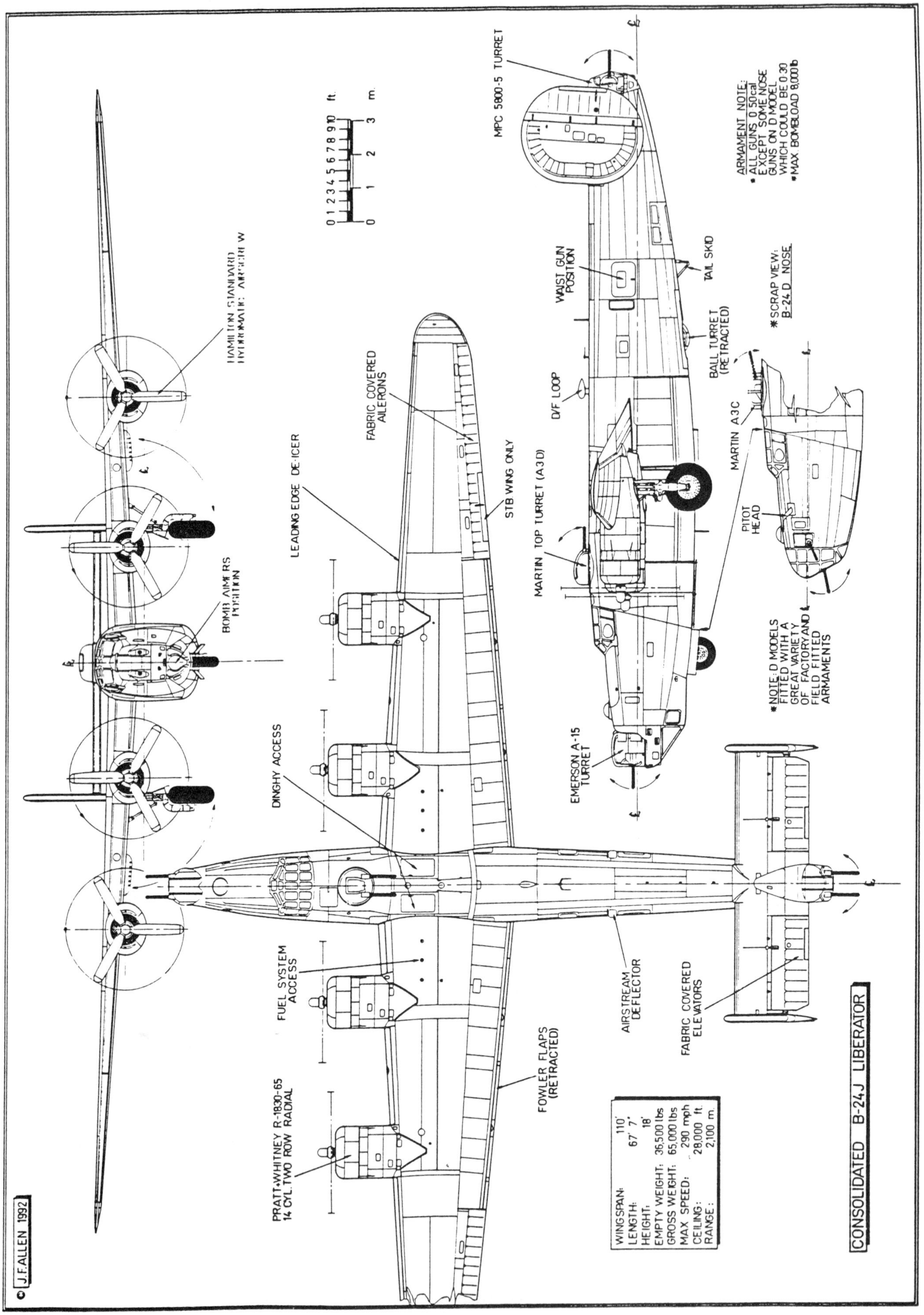
© J.F.ALLEN 1992
CONSOLIDATED B-24J LIBERATOR
WINGSPAN: 110'
LENGTH: 67' 7"
HEIGHT: 18'
EMPTY WEIGHT: 35,500 lbs
GROSS WEIGHT: 65,000 lbs
MAX SPEED: 290 mph
CEILING: 28,000 ft.
RANGE: 2,100 m.
PRATT+WHITNEY R-1830-65 14 CYL. TWO ROW RADIAL
FUEL SYSTEM ACCESS
DINGHY ACCESS
LEADING EDGE DE-ICER
FABRIC COVERED AILERONS
STB WING ONLY
FOWLER FLAPS (RETRACTED)
AIRSTREAM DEFLECTOR
FABRIC COVERED ELEVATORS
BOMB AIMERS POSITION
HAMILTON STANDARD HYDROMATIC AIRSCREW
0 1 2 3 4 5 6 7 8 9 10 ft.
0 1 2 3 m.
EMERSON A-15 TURRET
MARTIN TOP TURRET (A3D)
D/F LOOP
WAIST GUN POSITION
BALL TURRET (RETRACTED)
TAIL SKID
MPC 5800-5 TURRET
ARMAMENT NOTE:
* ALL GUNS 0.50cal EXCEPT SOME NOSE GUNS ON D MODEL WHICH COULD BE 0.30
* MAX BOMBLOAD 8,000lb
* SCRAP VIEW: B-24 D NOSE
MARTIN A3C
PITOT HEAD
* NOTE: D MODELS FITTED WITH A GREAT VARIETY OF FACTORY AND FIELD FITTED ARMAMENTS

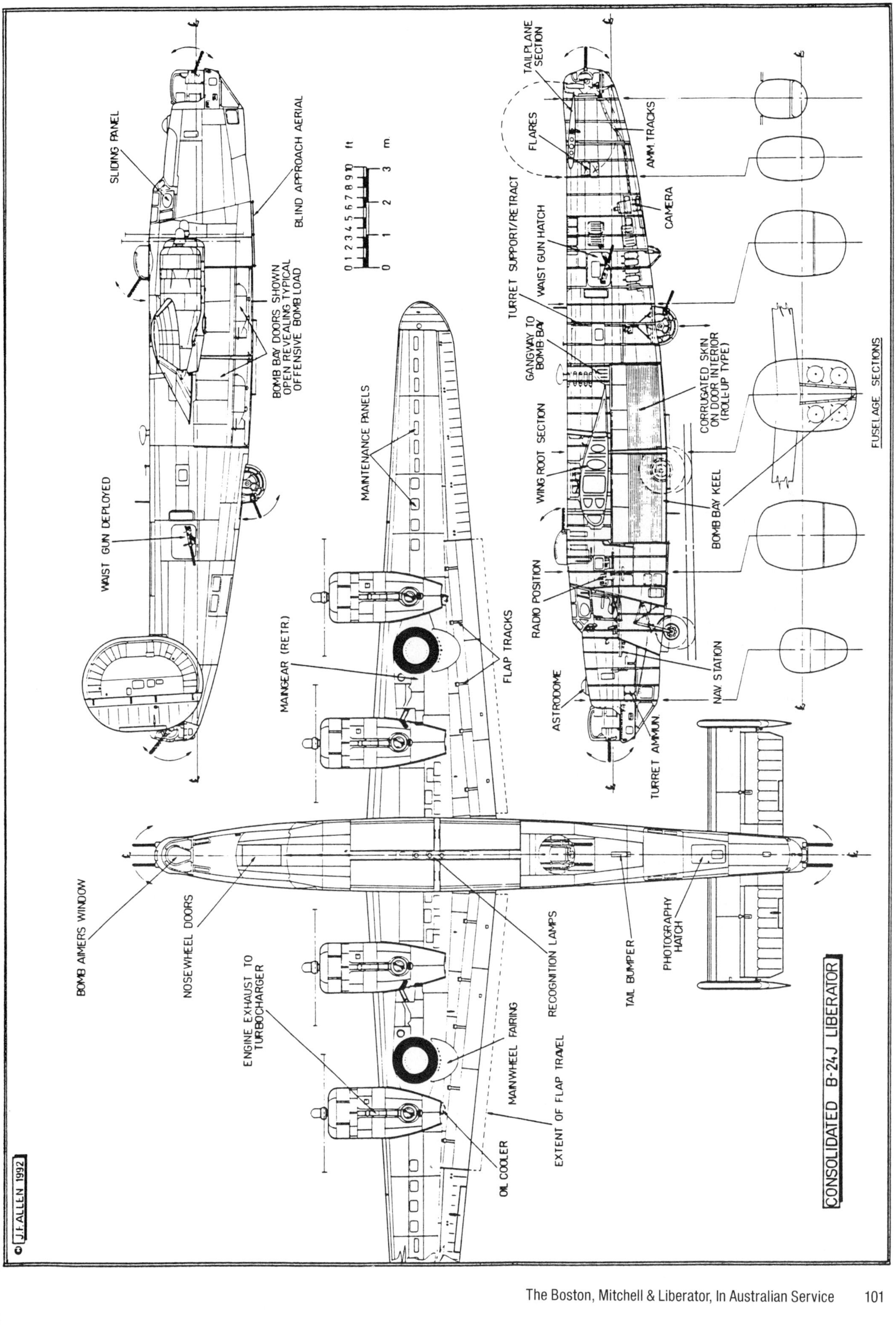

CONSOLIDATED B-24J LIBERATOR
© J.F. ALLEN 1992
SLIDING PANEL
BLIND APPROACH AERIAL
BOMB BAY DOORS SHOWN OPEN REVEALING TYPICAL OFFENSIVE BOMB LOAD
WAIST GUN DEPLOYED
0 1 2 3 4 5 6 7 8 9 10 ft
0 1 2 3 m
MAINTENANCE PANELS
MAINGEAR (RETR.)
FLAP TRACKS
BOMB AIMERS WINDOW
NOSEWHEEL DOORS
ENGINE EXHAUST TO TURBOCHARGER
RECOGNITION LAMPS
MAINWHEEL FAIRING
EXTENT OF FLAP TRAVEL
OIL COOLER
TAIL BUMPER
PHOTOGRAPHY HATCH
TAILPLANE SECTION
FLARES
AMM. TRACKS
CAMERA
TURRET SUPPORT/RETRACT
WAIST GUN HATCH
GANGWAY TO BOMB-BAY
CORRUGATED SKIN ON DOOR INTERIOR (ROLL-UP TYPE)
WING ROOT SECTION
BOMB BAY KEEL
RADIO POSITION
NAV. STATION
ASTRODOME
TAIL AMMUN.
TURRET AMMUN.
FUSELAGE SECTIONS

The final mass produced Liberators were the B-24L (1,667 built) and B-24M (2,593 built). Both were similar to late model B-24Js, the 'L' featuring the 'Stinger' tail guns and the 'M' a new lightweight tail turret. To the end, the Liberator's defensive armament was being fiddled with! A B-24M was the 6,725th and last Liberator to come off Consolidated's San Diego production line.

Odds and Sods

There were several one-off experimental Liberators, including the XB-41 gunship intended for bomber escort duties and armed with no fewer than fourteen 0.50in machine guns with 12,420 rounds of ammunition. Flown in February 1943, the XB-51 was converted from a B-24D and featured five twin gun turrets (chin, ventral, tail and two mid-uppers) plus two twin waist mountings. The gunship concept was found wanting, so the XB-41 was converted to TB-24D configuration as an instructional airframe. Five B-24Ds were converted for specialist training duties as AT-22s and were later redesignated TB-24s.

One B-24 'minor' variant which was built in substantial numbers was the C-87 Liberator Express transport. Considering the early and successful use of the Liberator in the transport role this was an obvious version to produce in numbers. Two hundred and ninety-one were built including 25 for the RAF and six C-87A deluxe passenger transports with sleeper accommodation. Power was provided by turbocharged R-1830-43 engines. The C-87B was a proposed armed variant and the C-87C had a single fin and rudder and lengthened for-

A US Navy PB4Y-1 Liberator, mainly based on the B-24D but in this case fitted with a nose turret. Again, the slender Davis wing is emphasised.

CONSOLIDATED B-24J LIBERATOR
Powerplants: *Four Pratt & Whitney R-1830-65 Twin Wasp 14 cylinder twin row air cooled radial engines with single stage engine driven superchargers and General Electric B-22 exhaust driven turbo-superchargers. Rated at 1,200hp for takeoff, 1,200hp maintained up to 31,800 feet at Military Rating. Hamilton Standard Hydromatic three-bladed constant speed and feathering propellers of 11ft 7in (3.53m) diameter. Normal fuel capacity of 1,978imp gal (8,992 l) in 12 self sealing cells in centre section between spars; maximum fuel capacity 3,009imp gal (13,679 l) if additional outer wing and bomb bay tanks fitted.*
Dimensions: *Wing span 110ft 0in (33.53m); length 67ft 2in (20.47m); height 18ft 0in (5.48m); wing area 1,048sq ft (97.36sq m); wing loading (normal max weight) 62.0lb/sq ft; power loading (normal max weight) 13.5lb/hp.*
Weights: *Empty 38,000lb (17,237kg); normal max takeoff 65,000lb (29,484kg); max overload 71,200lb (32,296kg).*
Armament: *Typical bomb load 5,000lb (2,268kg); normal max bomb load 8,000lb (3,629kg); max over short ranges 12,800lb (5,806kg). Defensive armament typically ten 0.50in Browning machine guns in four twin gun turrets (Consolidated A-6 or Emerson A-6 nose, Sperry Ball ventral, Martin A3 upper, Consolidated A-6 tail) and single manually operated waist mounts; total ammunition 4,716 rounds.*
Performance: *Max speed 300mph (483km/h) at 30,000ft and light weight; max continuous 278mph (447km/h) at 25,000ft; typical cruise 237mph (382km/h) at 25,000ft; initial climb 1,025ft (312m)/min; service ceiling 28,000ft (8,534m); range with 5,000lb bomb load 1,700 miles (2,736km) at 232mph; range with 8,000lb bomb load 1,540 miles (2,478km); max range (normal fuel) 2,100 miles (3,379km).*

ward fuselage. This version was designated Liberator C.IX by the RAF and RY-3 by the US Navy. Winston Churchill's LB-30A "Commando" was modified to C.IX standards in 1944 (including the single fin) but retained its mechanically supercharged R-1830-33 engines.

The basic C-87 (RAF Liberator C.VII) was developed from the B-24D with turrets removed and faired, a floor installed in the bomb bay and windows and freight doors cut into the fuselage. Liberator transports played an important role in the war years and immediate postwar period and many were converted in Great Britain and Canada from surplus bombers.

The F-7 was a photographic reconnaissance conversion of the Liberator, the first based on the B-24D and the remaining 214 (F-7A and B) on the B-24J. The F-7A carried three nose and three bomb bay cameras and the F-7B had six cameras installed in the bomb bay. Bombing equipment was removed but defensive armament retained.

Several different Liberator variants formed the basis of the 218 C-109 tankers, designed to transport fuel to various theatres of operation, mainly in support of the B-29 Superfortresses operating from China. Stripped of all military equipment and armour, the C-109s could carry a total fuel load of 2,400imp gal (10,900 l).

Liberators with a single fin and rudder assembly were investigated as early as 1943 with the conversion of a B-24D to this configuration along with the installation of more powerful 1,450hp R-1830-75 engines as the XB-24K. It wasn't until 1944 that the decision was made to fit all future Liberators with this feature (and the more powerful engines) but by the time the production model, the B-24N, became available, manufacture of the Liberator generally was about to cease. As a result, only seven B-24Ns were built but these featured yet another variation on the Liberator's defensive armament in the form of an Erco ball turret in the nose and a modified rear turret which resembled that which was installed on the Boeing B-17 Fortress.

The concept of the single finned Liberator was able to continue with the US Navy's PB4Y-2 Privateer patrol bomber, 739 examples of which were delivered between March 1944 and October 1945. Powered by unsupercharged 1,350hp R-1830-94 engines, the Privateer retained the standard Liberator wing mated to a fuselage stretched by seven feet (2.13m) and a very tall fin and rudder assembly. Six turrets each housed twin 0.50in machine guns – Erco nose turret, two Erco waist 'blister' turrets, two Martin upper turrets, and Consolidated tail turret. As the final expression of the extremely important but often understated Consolidated Model 32/B-24 Liberator series, the Privateer remained in service with the US Coast Guard as late as the mid 1960s.

The final expression of the Liberator series, the US Navy's PB4Y-2 Privateer patrol bomber of 1944. This combined the standard Liberator wing with a lengthened fuselage, a single fin and rudder and more powerful engines.

The Liberator was the first and most numerous heavy bomber to be taken on charge by the Royal Australian Air Force, a total of 287 of them in four different versions being delivered to the RAAF between February 1944 and August 1945. It was also the RAAF's first four engined bomber. Carrying the serial number prefix A72, the Liberators served Nos 12, 21, 23, 24, 25, 99 and 102 Squadrons, their wartime activities including playing an important role in the Borneo campaign. Postwar, the Liberator served until February 1948 when the three squadrons based at Amberley, Queensland, were disbanded.

Early Connection

Australia's association with the Liberator began well before the type was introduced to RAAF squadron service in 1944. With the Japanese attack on Pearl Harbour on 7 December 1941 rapid action was instigated by the United States Army Air Force to put it on a war footing, including the diversion of 75 Liberator IIs intended for the Royal Air Force to US service. Redesignated LB-30s and retaining their RAF serial numbers, 15 of these aircraft (along with 65 Boeing B-17Es) were ordered to the Pacific area, specifically to reinforce Java in the Netherlands East Indies (now Indonesia) in the face of the rapidly advancing Japanese. Their unit was the 11th Bomb Squadron of the 7th Bombardment Group and before departure the aircraft were hurriedly modified to incorporate a Martin dorsal turret housing two 0.50in machine guns and a pair of similar guns in a handheld mount at the tail.

One Liberator was lost in the USA before delivery, another ditched *en route* and a third was delayed while accident damage sustained in the USA was repaired. By the time this particular Liberator (AL573) arrived in the area, Java had fallen to the Japanese.

The Java campaign was short and costly, lasting just over a month in January and February 1942. During that time the Liberators participated in numerous raids against Japanese targets including an aircraft carrier, targets in the Celebes and Sumatra and against shipping during the Japanese invasion of Bali.

By late February the position on Java was untenable and the five surviving LB-30s were used for the evacuation of American and Dutch personnel from Java to Darwin and Broome, in conjunction with five USAAF Ferry Command B-24As and other types. Darwin had previously witnessed the end of one of the LB-30s (AL521) during the first air raids on the city on 19 February. Broome was itself raided on the morning of 3 March 1942, the attack claiming 23 USAAF, USN, RAF, RAAF, Dutch and civilian aircraft including two USAAF B-24A Liberators. One of the now only three remaining LB-30s, AL515, had been in Broome the previous day, but with the arrival of a Japanese reconnaissance aircraft, quickly departed.

The three LB-30s which survived all this were AL508, AL515 and AL570 and they were augmented at the end of February by the late arriving AL573 which had been delivered to Sydney but arrived at Broome the day after the raid.

Some reorganisation of the American forces followed and the LB-30s were assigned to the 28th Squadron of the USAAF's 19th Bomb Group (which incorporated surviving elements of the decimated 7th), fitted with radar and cameras and sent to Cunderdin WA where they patrolled their particular section of the Indian Ocean. From August 1942 the LB-30s operated from Townsville Qld, flying long range reconnaissance missions including the first over the Japanese island of Truk and some of the first sorties over Guadalcanal.

By September 1942 only two of the LB-30s remained. AL508 had been destroyed after crashing on takeoff at Laverton Vic the previous May, while

B-24J Liberator A72-34 of No 24 Squadron, the first RAAF operational Liberator squadron. This aircraft was part of the first batch of 38 B-24Js delivered to the RAAF between May and October 1944. (via Neil Mackenzie)

US Army Liberator LB-30 AL570, one of the survivors of the Java campaign. This aircraft went on to fly with several USAAF units in Australia and with No 36 Squadron RAAF. The photograph was taken at Maylands, Perth in March 1942, immediately after the fall of Java. (via Geoff Goodall/Bob Livingstone)

AL515 suffered a collapsed undercarriage on landing at Milne Bay in August. The aircraft was strafed and destroyed by Japanese 'Zero' fighters before it could be recovered and repaired.

This left AL570 and AL573, both of which remained in Australia until 1944 after having been converted to C-87 transport configuration. Their subsequent careers in Australia will be noted in the following chapter, including – in the case of AL570 – a stint on the official strength of the RAAF's No 36 Transport Squadron.

American Liberators Up North

The next USAAF Liberator association with Australia was a rather more successful and less hastily organised one. Liberators of the USAAF's 5th Air Force began moving into the Northern Territory in mid 1942 as part of the general build up of the time which saw Northern Australia become a very large base for Allied land, sea and air forces. The Liberators belonged to the USAAF's No 380 Bombardment Group which had its headquarters at Fenton, south of Darwin. The Group's four squadrons were divided between there and Long strip, from which they could fly to targets deep in Japanese held territory. The 380th BG's Liberators were joined by the 90th's similar aircraft late in 1942.

The importance of the Fenton Liberator base was not lost on the Japanese. On 30 June 1942, 27 'Betty' bombers and 23 escorting fighters bombed the airfield destroying three Liberators and damaging a further seven. The raid was intercepted by the three Spitfire squadrons of Darwin's 'Churchill' Wing but a misunderstood order resulted in the Spitfires which were assigned to attack the bombers having to defend themselves against the escorting fighters because the third Spitfire element had not taken on the fighters as instructed. Nevertheless, six Japanese bombers and two fighters were shot down.

The USAAF Liberators deployed to Nadzab in New Guinea in early 1944, at the same time the first of the RAAF's own aircraft were delivered. American Liberator units would shortly play an essential part in the training of RAAF crews on their first heavy bomber.

What is not generally known is that USAAF Liberator units operated under RAAF command and continued to do so even after the RAAF's Liberator squadrons were up and running. When that occurred, the American and Australian squadrons often operated in concert and provided the main offensive strength to the north of Australia.

With the deployment of the USAAF Liberators to the northern Australia/New Guinea area, came an extensive infrastructure including a large service and repair organisation (the 81st Air Depot Group at Archerfield and Eagle Farm, Brisbane) and a refresher training organisation. These factors – plus ready availability – made the Liberator the obvious choice as the RAAF's heavy bomber for the war in the Pacific.

Liberators for the RAAF

Earlier investigations into a bomber for the RAAF had begun in 1941 when proposals to manufacture such an aircraft in Australia were examined. The Avro Manchester was suggested as a possibility but the failure of its two Rolls-Royce Vulture engines to achieve reliability ended that idea. Instead, it was decided to concentrate Australian manufacturing effort on smaller aircraft such as the Beaufort and Beaufighter.

In late 1942 it was decided that Australian representatives be sent to the United Kingdom and United States to evaluate other types of aircraft suitable for local manufacture, including heavy bombers. The Australian Production Mission spent the first half of 1943 fulfilling its primary function of selecting high and low altitude fighters and a general reconnaissance bomber "for the specific duty of being able to carry out overseas flights of up to 2,000 nautical miles carrying the equipment and crew necessary for GR [general reconnaissance] work".

Among the party was the Commonwealth Aircraft Corporation's L J (later Sir Lawrence) Wackett, a man whose influence on Australian aircraft design and production was to span four decades.

Following his trip with the Australian Production Mission, Wackett prepared a lengthy report (covering not only aircraft but powerplants, armament and production techniques as well) which helped ensure the North American Mustang fighter would later be built in Australia, but also dealt with the question of which bomber would eventually be coming out of Australian factories.

Wackett dealt with the Liberator thus: "... it has become generally recognised that four engined aircraft are greatly to be preferred *(sic)* for overseas flights, and it has proven that the B-24 Liberator is capable of carrying out the duties of a GR aircraft, so as to effectively meet the Air Board Specification". Despite praising the Liberator, Wackett could find no

B-24D Liberator A72-6, one of the first group of aircraft delivered to the RAAF in February 1944 and used to form the early strength of No 7 Operational Training Unit. Of note is the nose turret, a feature of some later B-24Ds. This particular Liberator survived the war and was converted to components in 1947.

(right) RAAF Chief of Air Staff from 1942-52 Air Vice Marshall George Jones (left) photographed with General George C Kenney, the commander of all Allied Air Forces in the South West Pacific Area. Kenney was instrumental in arranging both the procuring of Liberators for the RAAF and for the training of RAAF aircrew with the USAAF's 5th Air Force.

reason to manufacture the Liberator in Australia because "... this aircraft is in very large scale production in several places and it does not appear justifiable to start out production in Australia at this late date".

Wackett was perfectly correct, of course, and the RAAF's Liberators were manufactured in the USA. He recommended the Lancaster for Australian production and in its improved Lincoln form, 73 were delivered to the RAAF, the first of them flying in March 1946. Perhaps ironically, the Lincoln replaced the in many ways more modern Liberator in RAAF service.

In late 1943, US General George C Kenney, the commander of all Allied Air Forces in the South West Pacific Area, decided that due to the vast numbers of

Group Captain Deryck Kingwell, commanding officer of No 82 Wing RAAF and a prominent figure in RAAF Liberator operations. He is sitting in B-24L A72-81. (via Neil Mackenzie)

Two shots showing activity at 7 OTU Tocumwal during 1945, with (according to the official caption) "Crews are assembling for final instructions from their commanding officer (Grp Capt A A Barlow) and flying instructor before taking off on a training flight". No 7 OTU was the RAAF's largest Liberator user and at its peak in early 1945 had 58 of them on strength. (via Neil Mackenzie)

Liberators being manufactured in the USA, and in order to move the USAAF's 380th Bombardment Group from the Northern Territory to New Guinea, the RAAF should form seven Liberator squadrons.

The original intention was that the RAAF's No 99 Squadron, which was scheduled to be formed in Queensland in March 1944 would be the first Liberator equipped unit, with 12 aircraft. This was changed when it was decided that the Vultee Vengeance dive bombers operated by Nos 21, 23 and 24 Squadrons in New Guinea were unsuitable for further operations and would be withdrawn to Australia. Instead, No 24 Squadron would be the first to re-equip with Liberators (in July 1944) followed by Nos 21 and 23 by the following October. Together, the three squadrons formed 82 Wing, RAAF.

It was realised from the outset that it would be difficult to find the number of experienced aircrew required to fly these aircraft at short notice, particularly since the demands of the European air offensive would make it almost impossible to divert RAAF heavy bomber crews from that part of the world. General Kenney therefore agreed to assist the RAAF with both the procurement of aircraft and the training of crews so as to have the new squadrons operational as soon as possible. Accordingly, he arranged with the Australian Chief of Air Staff (AVM George Jones) for RAAF crews to be attached to the US 5th Air Force for conversion training and operational experience.

Negotiations with the USAAF were conducted by two RAAF officers, Sqdn Ldr J B Hampshire and Flt Lt F X McMahon at Port Moresby in December 1943. It was agreed that when attached to the 5th Air Force, RAAF crews would be trained as such, and not as members of American crews. The complete RAAF crews would pass through a Heavy Bomber Replacement and Training Centre, where they would undergo training on lines similar to an RAAF twin engined Operational Training Unit (OTU) course. It was planned that the first five crews would be converted to multi-engined heavy aircraft by May 1944. The course was scheduled to take a month to complete and include some 100 flying hours comprising bombing, formation and air gunnery.

Conversion training of the initial five crews began at the end of December 1943 at Charters Towers (Queensland). Subsequently, a further five crews completed their conversion courses at Jackson Field, Port Moresby in March 1944, and Course 3, comprising eight crews, completed training in April 1944. By then, most of the initial batch of 12 B-24D Liberators earmarked for the RAAF had arrived and the first of large quantities of the more capable B-24J and M variants were close to delivery. The first B-24Js arrived in May 1944.

Training and Squadron Formation

With the completion of training of the first five RAAF Liberator crews at the end of January 1944, they were assigned to the USAAF's 43rd Bombardment Group at Nadzab for combat training. It was recommended by USAAF authorities that when trained, the RAAF crews should be formed into a separate flight attached to the 380th BG for tactical operations and that this flight would remain intact until the formation of the first RAAF Liberator squadron, which would be stationed in the Darwin area.

The newly formed Directorate of War Training at RAAF Headquarters was appointed the co-ordinating body for conversion training with the American OTU and was also made responsible for the formation and war training of all new squadrons until these were handed over to RAAF Command.

General Kenney decided that in order to expedite the formation of No 24 Squadron, it would be established progressively in conjunction with the 380th BG and the first Liberators assigned to 24 Squadron would form an RAAF element within the Group, which would be gradually built up until one complete squadron was formed. Once the squadron had sufficient aircraft and crews to carry out independent operations, the 380th BG's responsibility for it would end.

A conference on aircraft allocations held in Washington DC during January 1944 determined that no new aircraft would be made available for the seven planned RAAF Liberator squadrons for the next six months. If aircraft became available, it was hoped to form these new squadrons at a rate of one per month from July 1944. To help things along, the 5th Air Force agreed to supply the RAAF with 15 used Liberators in February 1944 with the intention of using these aircraft to form the first squadron as quickly as possible.

But the RAAF decided that these Liberators (13 as it turned out; (12 B-24Ds and one B-24J, serial numbers A72-1 to -13) were suitable only for training, and that the formation of the first B-24 flight would have to be deferred until the arrival of new aircraft from overseas. These Liberators (B-24Js initially) began arriving in May and were allocated serial numbers from A72-31 onwards.

In March 1944 the Australian Chief of Air Staff requested the withdrawal of RAAF crews engaged in New Guinea with the 380th BG in order to start forming RAAF Liberator squadrons. General Kenney agreed to the release of five crews when their tours of duty ended in April, and because of unsuccessful attempts to obtain conversion training

An early RAAF Liberator casualty was B-24L A72-123, which crashed on landing at Tocumwal in January 1945, just two months after it was delivered. (RAAF Historical)

for RAAF aircrews in the United Kingdom and United States, Kenney also agreed to increase the number of RAAF crews to be trained by the 5th Air Force from 25 to 80.

By May 1944 the B-24 flying training programme was in full swing, with the 30 day course at Jackson Field able to turn out five to ten complete crews per month. Advanced technical training equipment and experienced instructors were available and the RAAF was given full co-operation by the Americans. After completing the course, RAAF crews remained with the 380th BG for a further two months, and by June 1944 it was estimated that when sufficient aircraft became available for the formation of No 24 Squadron, 12 crews would be ready for posting.

7 OTU Formed

In the meantime, the RAAF's No 7 Operational Training Unit had been formed at Tocumwal in February 1944 to provide aircrew for the RAAF heavy bomber squadrons. 7 OTU's establishment was set at six Liberators from the initial batch along with two Vultee Vengeances and a single Airspeed Oxford. Considerable difficulties were encountered during 7 OTU's formation period and the training programme was delayed due to a lack of Liberator spares, a problem exacerbated by the fact that its aircraft had already seen active service with the USAAF and were in a generally poor state of repair. In addition, Tocumwal's buildings were well dispersed, living conditions in general were far from ideal and the station's administration was considered to be unwieldy and divided.

Despite this, the first Liberator course at 7 OTU was completed by the beginning of June 1944 and four crews were ready to be posted to No 24 Squadron. Including the 12 crews which had trained with the Americans, a total of 16 Liberator crews were ready for 24 Squadron when they required.

RAAF LIBERATORS
SUMMARY OF SERIALS AND DELIVERIES

Serial Nos	Model	Qty	Delivery
A72-1/9	B-24D	9	02/44
A72-10	B-24D	1	05/44
A72-11/12	B-24D	2	07/44
A72-13	B-24J	1	07/44
A72-31/68	B-24J	38	05/44-10/44
A72-69/142	B-24L	74	09/44-12/44
A72-143/148	B-24M	6	11/44-12/44
A72-149/157	B-24L	9	11/44-12/44
A72-158/198	B-24M	41	12/44-02/45
A72-300/405	B-24J	106	03/45-08/45

Liberators Arrive

By the end of May 1944, two new B-24J Liberators had arrived in Australia, four others were *en route* and eight additional aircraft were ready for ferrying from the USA. All 14 (A72-31 to -44) were in Australia by the end of June and a continuous flow of Liberators would follow. By the end of 1944 some 140 B-24J, L and M Liberators had been delivered to the RAAF. Of the 287 Liberators ultimately taken on RAAF charge, 168 were allotted under the auspices of the Lend-Lease scheme.

Of the first 14 B-24Js, 12 were alloted to No 24 Squadron and after being received by No 3 Aircraft Depot (AD) and having LORAN navigation equipment and Norden bombsights installed, were flown to their intended area of operation.

By late June 1944, No 24 Squadron's headquarters had moved from Lowood Qld to Manubullo in the Northern Territory, and by early July the squadron was fully operational under the command of the now Wing Commander J B Hampshire. No 24 Squadron moved to Fenton the following September, continuing the tradition of Liberator operations from that airfield.

It had been intended that No 21 Squadron would re-equip with Liberators at Leyburn, Queensland, during July 1944 but a shortage of aircraft made it necessary for the squadron to build up gradually and the date it would receive its Liberators was postponed until September. It was also planned that the squadron would move to Fenton at the beginning of October, but shortages of equipment and aircraft in conjunction with maintenance and transport difficulties caused further delays. It wasn't until January 1945 that No 21 Squadron was fully operational, with its headquarters at Fenton.

Similar problems also delayed No 23 Squadron's rise to full operational capability – despite the large number of Liberators which had been received – until April 1945 when it moved from Leyburn to Long, NT, under the command of Wng Cdr R A Dunne. All three squadrons were part of the RAAF's No 82 Wing, commanded by Grp Capt D W Kingwell, a 27 years old 'veteran' of RAAF Hudson operations in New Guinea.

Training Revised

Formed in June 1944, the Far East Air Force (established to control the USAAF's 5th and 13th Air Forces and commanded by General Kenney who would remain commander of Allied Air Forces) took over the combat training of US aircrew from the 5th Air Force. It was therefore

B-24J Liberator being prepared for delivery to No 24 Squadron in September 1944. This aircraft was lost in action over Balikpapan in July 1945. (via Neil Mackenzie)

A72-76, a B-24L Liberator, in the process of being hooked up for a tow at Tocumwal. (via Neil Mackenzie)

decided that the American Replacement and Training Centre would move from Jackson to Nadzab and would become a training centre for not only heavy bomber but also fighter and medium bomber crews. The training would comprise not the conversion from one type of aircraft to another but refresher training for new crews arriving in the area.

This change meant altering the training arrangements of RAAF Liberator crews, who would now receive eight weeks training (120 flying hours) at No 7 OTU and then go to Nadzab for a final one month's training (40 to 45 training hours) at the Combat and replacement Training Centre. The RAAF contributed about 140 ground staff to assist in the maintenance duties at Nadzab, and these men, along with the RAAF aircrews under training were formed into a separate unit called the Heavy Bomber Replacement Training Unit (HBRTU).

The Unit formed at Strathpine Qld in September 1944 and began training at Nadzab the following November under the command of Wng Cdr J E Handbury. The American training centre provided Liberators and both flying and ground instructors for the crews and the unit was required to give advanced training to up to 22 crews per month. The crews flew several combat missions over Wewak and Rabaul as part of their training.

In December 1944, arrangements were made to withdraw the HBRTU from Nadzab and in February 1945 the unit disbanded at Strathpine. Between the inception of the Liberator training scheme in December 1943 and then, a total of 103 RAAF heavy bomber crews had been trained in co-operation with the USAAF.

The operational histories of the three Liberator squadrons comprising No 82 Wing and the other RAAF squadrons and units which operated the aircraft will be discussed in the following chapter.

B-24J Liberator A72-319 from the final batch of 106 aircraft delivered between March and August 1945. The USAAF serial number (44-28082 in abbreviated form) is visible on the fin.

LIBERATOR CREWS- HEAVY BOMBER REPLACEMENT TRAINING UNIT.

Interview with F/Lt. D WHiggins, Liaison Officer, and F/O D R COSSINS, Adjutant, by N.G. Asst.N. at NADZAB on 18th Nov 44.

F/Lt. Higgins stated:

We are now on course 8 at the unit. All trained crews from previous courses have gone to the 380th Bomb Group at Darwin After going to Darwin they do 10 strikes under the guidance of the Americans.Then their training is completed and they are handed over to our Liberator squadrons.. The firt courses commenced training at Strathpine under the control of the Americans and a liaison officer was appointe to look after RAAF interests. The first liaison officer was P/O Mc Pherson add I took over from him in July 44. The unit formed at Strathpine on the 1st September 1944 as a separate RAAF unit to administer the crews who were being trained by the Americans. Prior to that there had been no RAAF organisation the crews being merely infiltrated with the Americans. The rear party of HBRTU was still at Strahtpine although the advance party arrived at Nadzab on October 8th, 1944. The purpose of the unit is purely administrative all training add opeations being controlled by the Americans.

The original instructions relating to Liberator training were that five crews were posted to 1 RPP for disposal by 5 RPSO to commence training with the Americans at Charters Towers. The course under the command of S/Ldr. J.B. Hampshire was to move to Port Moresby with the Americans where they were to be posted to 9 O.G. and assigned for duty with the 5th Air Force.

The first course commenced training at Charters Towers on 6th Feb 1944. and then went to Nadzab where they were with the 65 bombardment squadronnof the 43rd bomb group. Prior to that they had done 2 or 3 weeks at C. Towers. From Nadzab they did operations with the Americans. The first five crew pilots were S/Ldr. J.B. Hampshire, S/Ldr. JT O'Brien S/L W J Rehfisch F/Lt. G A F Jacques and S/Ldr. RC W Rayson. Raysons crew crashed while taking off and became casualties. After the first course had been completed the crews were posted back to the mainland for Air Board disposal, some of them, I believe, going to American for ferry duties.

No. 2 Course began at Jackson Field Port Moresby on the 12th February. The pilots were F/Lt. R F Oberhue (course captain), F/Lt J C A Napier. F/Lt. C E R Parsons, F/Lt. I A McCoombe and F/O WJ Stutt On completion of their training these crews were posted to the 380th Bomb Group at Darwin They left Nadzab on the 18th March 44

The third course under S/Ldr. L W Manning commenced on the 15th March and completed training on the 27th April. No. 4 began part training on the 30th April 44 commenced full training on the 5th May add finished on the 9th June. G/Cpt. D W Kingwell was course captain.

No.5 Course Began 19th June Finished 20th July Captain S/Ldr. R EBELL.

No. 6 Course began 11th August completed 25th September. Captain, S/Ldr. RT Cupper. This course came from Port Moresby to Nadzab on the 11th September to complete their training.

No. 7 course under S/Ldr. G.L. Brown began training on the 25th September This course was broken, three crews finishing on the 1st November and the other 9 on the 5th November.

No. 8 Course is the first course under the new H B R T U set-up. W/Cdr J Chandbury is now the commanding officer of the unit and the senior member of the course is W/Cdr. N.G. Hemsworth.

On the 12th August 1944 Northern Command was advised by Air Board that RAAF HQ had approved of a general plan for a RAAF unit within the FEAF organisation at Nadzab Establishment action had been taken to provide a self-contained RAAF unit underthe command of a GD W/Cdr. The purpose of this unit was to organise and control all apsects of RAAF domestic administration, discipline etc within the general organisation of FEAF A maximum of 22 crews had been posted and action was being taken to post 150 technical personnel.

Before coming up here the crews are doing about 100 hours at Tocumwal The navigators do a radio navigation course and the pilots learn flying in tropical conditions The bomb aimers get experience on the Norden bomb site and the working of the turrets. The previous course did strikes over Wewak and No. 8 Course is now doing strikes over Rabaul. There are 11 members of the RAAF crew, pil t, second pilot, bombardier, navigator, flight engineer, wireless operator, asst. wireless operation, 4 air gunners.

Between the 5th and 16th February No. 1 course did two or three strikes. Rayson when taking off at 0817 hours on the 7th Mar 44 crashed at the end of the strip with 8000 lbs bombs and the a/c disintegrated. All members of the crew were injured. This has been the only casualty to date.

W/Cdr Hanbury explained that the unit,which was a part of the American Combined Training Replacement Centre at Nadzab, was engaged in the advanced training of RAAF Liberator crews.The CRTC was theAmerican equivalent of the Australian O.T.O. and the underlying idea was to give heavy bomber crews experience under conditions as close as possible to actual combat and experience of flying in tropical weather.The heavy bombers did not go over targets where fighters might be met.Strikes included American Bostons which carried out strafing attacks and B 25's which both bombed and strafed.In attacking targets in the Wewak area bombing was carried out at aheight of from 4,000 to 5,000 ft, and in the Rabaul area, where there was a good deal of a/a/ opposition still, from 14000 - 20,000 ft.

HBRTU crews were instructed by experienced Americans and were using American machines, of which the unit had 10 - a number which it was expected would be built up to 12.RAAF personnel serviced all aircraft being used by AustraliansThese airmen had their own camp set up within the American servicing group and the aircrew members had theirown section of the CRTC camp area.The unit had its own engineer officer who was not F/Lt J EJ Summers.

A proportion of Tocumwal trained crews was included in each course and on completion of the five weeks course the trained crews were sent to Darwin where they became part of the 380th Bombardment Group. In that capacity they were required to carry out from 10 to 15 strikes or 100 flying hours on operations before they were sent finall as replacements -y/to a RAAFSquadron for operations.

Crews on course at Nadzab had been making attacks on Wewak, and and No 8 course was going to strike against Rabaul targets, with the possibility that future targets of the unit would include those specially selected in support of A I F operations.

There were about 140 ground staff personnel comprising the units servicing parties.

The 'Belle' Story

The previous chapter made mention of the USAAF LB-30 Liberators which participated in the Java campaign in early 1942 and the conversion of the sole survivor of that group (AL570) and another which arrived shortly after Java was lost to Japanese (AL573) to C-87 transport configuration later in the same year. These Liberators both continued their operational careers with an Australian connection, the former eventually becoming the first Liberator to appear on the strength of an RAAF Squadron.

Following conversion to freighters by the addition of a large cargo door to the port rear fuselage and a nose hatch, AL570 was given the name *Nipponese Nipper* and the Directorate of Air Transport callsign VHCBL. AL573 became VHCBM (and was named *Kangaroo Kate* on one side of the fuselage and *Gopher Gus* on the other!) and both aircraft were assigned to the 21st Troop Carrier Squadron of the USAAF's 374th Troop Carrier Group.

AL573 was transferred to the 317th TCG at Townsville in March 1943 but had been put to pasture in a corner of Garbutt strip by the following September. In 1944 the Liberator was "liberated" by Consairway, a division of Consolidated Aircraft which operated under contract to the USAAF providing transport services between Australia, the USA and in the Pacific region. After receiving sufficient maintenance to get her airworthy again, AL573 was ferried back to the USA for a full overhaul and return to service with the Consairway fleet.

Nipponese Nipper (AL570/VHCBL) was transferred to the 317th TCG in May 1943 and joined its stablemate in the service of the 39th and 46th Troop Carrier Squadrons of that Group and then the USAAF 5th Air Force Service Command at Townsville. But by March 1944 this Liberator was also no longer required and was left parked at Archerfield aerodrome.

The Royal Australian Air Force's No 36 (Transport) Squadron was a regular visitor to Archerfield during this period, flying its C-47 Dakotas from Garbutt (Townsville) to New Guinea and points on the Australian mainland. Mindful of the extra carrying capacity offered by the Liberator and the fact that nobody seemed to want it, enquiries were made by the Australians as to its 'ownership' and availability. They were told by the Americans that the Liberator was theirs if they could get it up and running as it was considered unreliable, cantankerous and of no use to anyone.

They did, and renamed *Belle*, the old Liberator was taken on the strength of No

Liberator LB-30 AL570/VHCBL – "Belle" – at Higgins, Cape York Peninsula on 24 May 1944, a day fraught with difficulty as No 3 engine would not start. (Bob Livingstone)

A Liberator of No 21 Squadron RAAF in action in the South West Pacific Area. Note the 'roller shutter' bomb doors in the open position.

36 Squadron RAAF on 19 April 1944. Remembering that this was a very unofficial and temporary transfer, the aircraft carried USAAF markings during its time with the Australian squadron and never received an RAAF serial number – AL570 was retained. Three pilots were those which almost exclusively flew *Belle* during her time in Australian service; Sqdn Ldr Harry Purvis, Sqdn Ldr Ron Cornfoot and Flt Lt John Bonnington.

Purvis and Cornfoot were on the flight deck when *Belle* recorded the first incident of her RAAF career, just one week after entering service. On this occasion, the Liberator's nosewheel collapsed on landing at Ward's Strip at Port Moresby after a flight from Townsville. Normally, such an incident would not cause much of a stir, except on this occasion the payload of detonators and explosives in the cabin was augmented by a reputed 12 cases of gin in the nose locker, a somewhat unauthorised cargo.

It should be noted that the alcohol was in no way intended for any black market dealings. It was for some of the squadrons in New Guinea whose personnel were paying through the nose for black market liquor while this batch was legally purchased at much lower cost in Townsville.

Being in the nose of the aircraft during a nosewheel collapse, the cases of gin began to fly out of the aircraft as it sustained ever greater damage whilst scraping along the runway. When the aircraft stopped and the crew went looking for the missing bottles, all had disappeared, very quickly purloined by spectators lining the runway!

Naturally, Harry Purvis was required to explain his cargo to the Air Board, which was sympathetic and understanding, although the incident did inspire an enquiry into illegal liquor trafficking generally, with unfortunate results for some.

Belle was temporarily patched up and flown (nosewheel down) back to Queensland where repairs were undertaken by No 13 Air Repair Depot at Graddon near Townsville. By the third week of May she was back in service but only for a very short time.

AL570/VHCBL's subsequent career with the RAAF is a little vague. What is definitely known is that the RAAF's first Liberator was returned to the USAAF on 8 August 1944 and following the dissolution of the Directorate of Air Transport in early October, was ferried back to the USA a week or so after that.

B-24M Liberator A72-146 of 7 OTU and a P-40 Kittyhawk practice fighter affiliation over Tacoma NSW in 1945. (via Neil Mackenzie)

No 7 OTU

The early training and operational experience of RAAF Liberator crews was discussed in the previous chapter, and part of that training was the formation of the RAAF's No 7 Operational Training Unit (OTU) at Tocumwal, New South Wales. Its role as the RAAF's heavy bomber training unit was a vitally important one and at its peak in December 1944 it was the largest training unit in the RAAF with some 2,000 personnel on strength along with no fewer than 58 Liberators plus five Vultee Vengeances and an Airspeed Oxford.

7 OTU's first two B-24D Liberators (A72-7 and A72-8) were on strength by the end of February 1944 and by the middle of April A72-4, -3, -5 and -6 were also in service. At the same time, No 1 Liberator Course began flying and was completed by early June.

Deliveries of new Liberators continued throughout 1944 with 12 in service by late August, 17 by the end of September, 39 a month later and 46 by the end of November. Numbers remained high throughout the first half of 1945 (52-54 being typical) but naturally declined as the war eventually came to an end. By late November 1945 No 7 OTU's total strength was 38 Liberators, three Vengeances, three P-40 Kittyhawks, one Oxford and one Lancaster. The OTU was disbanded in early December 1945.

7 OTU suffered its fair share of accidents during its lifetime and casualties included several of the first batch of B-24Ds it acquired. Among them was A72-4 which was lost in unusual circumstances when it caught alight on the night of 2 October 1944 while electri-

cians were working on it. The subsequent fire destroyed the hangar in which the aircraft was parked and badly damaged it and Liberator A72-1. A72-10 was slightly damaged and survived the war.

24 Squadron RAAF

The first of the five former RAAF Vultee Vengeance squadrons to convert to the Liberator, No 24 Squadron's New Guinea detachment was withdrawn from operations in March 1944 and its personnel returned to Lowood in Queensland where the squadron was to be reformed as a heavy bomber squadron to be based at Manbulloo near Katherine in the Northern Territory. The first five Liberators arrived at Lowood in early June and through that month more aircraft were delivered and the squadron continued its move to its new base.

24 Squadron flew its first operational mission with Liberators on 6 July 1944 (a supply drop in western New Guinea) while the first combat mission was flown in early August when Flg Off M D Drecker made three bombing and strafing attacks on a 1,500 ton vessel near the Banda Islands. The squadron flew a total of 53

No 7 Operational Training Unit was the RAAF's largest user of Liberators. Illustrated here are B-24J A72-35 (top) and B-25L A72-126. 7 OTU Liberators carried their serial number suffix on the nose, in large yellow (with black border) figures. (via Neil Mackenzie/Mike Kerr)

missions against targets in and around the islands to the north of Australia during August.

In early September No 24 Squadron moved north from Manbulloo to Fenton Strip, closer to Darwin and the home of the USAAF's 380th Bomb Group, also equipped with Liberators. From Fenton, the squadron flew missions to targets in Java, the Celebes and Borneo and supported Allied operations in the Leyte Gulf area of the Philippines. Some missions were flown in conjunction with USAAF 380th Bomb Group Liberators, the unit under whose control No 24 Squadron operated until January 1945.

From that date, No 24 Squadron (along with the newly operational No 21 and the forthcoming No 23) would operate as part of the RAAF's No 82 Wing at Fenton. It remained based in the Northern Territory until June 1945 when the Wing moved to the large Allied base on Morotai Island in the Halmaheras group. From there, targets on Borneo, Celebes, the Philippines and other islands in the region were attacked. The squadron's work rate from Morotai was about double that which had been achieved from Fenton, with 113 missions being flown in June 1945 and 109 in July.

With the cessation of hostilities on 15 August, the number of missions flown was reduced to 43 and the squadron's activities revolved around shipping escort, transport and general patrol duties. No 24 Squadron began moving out of Morotai on 7 December 1945 – the fourth anniversary of the start of the war against Japan – and headed back to Australia to its new base at Tocumwal NSW. Transport duties between there and Morotai continued into 1946 until orders were received to cease operations at the end of March, leading to the squadron being disbanded in May.

Tocumwal became a 'graveyard' for the RAAF's Liberators and other types and after withdrawal from service, most languished there until sold for scrap in 1952.

Against Enemy Shipping

The following is a description of a successful operation carried out by five of No 24 Squadron's Liberators (A72-63, 70, 73, 88 and 106) on 2 December 1944 with orders "to search for and destroy shipping". The author of this originally handwritten piece is not known and it

Group Captain Deryck Kingwell taxies 24 Squadron's A72-81 (top) in after leading one of two strikes on the Kali Konto (Java) hydro-electric power station in early February 1945. In the lower shot, Flt Lt Kirkwood's aircraft is also taxying in. Operating from Truscott in Western Australia, the Liberators flew more than 1,000 miles to the target, which supplied power to the whole of east Java. (via Neil Mackenzie)

(above and opposite) 24 Squadron 'erks' hard at work on B-24J Liberator A72-40. These are official photographs issued by the Directorate of Public Relations and the supplied captions recognise the experience gained by Australian crews through their training with American units. (via Neil Mackenzie)

gives some insight into a part of 24 Squadron's duties at the time.

"Five Liberators took off from Fenton and proceeded to Truscott [on the northern tip of Western Australia] on the afternoon of 1st December 1944. At 012030Z [1st of the month, 2030 hours 'Zulu' or Greenwich Mean Time] the first duty [aircraft] was airborne from Truscott followed by the remaining duties at one minute intervals. They proceeded in formation along the designated tracks flying at 8,000 feet, until at 020221Z a convoy was sighted at position 03.30S/117.30E on course 120 degrees, speed 6-8 knots.

"The convoy consisted of one FTC estimated at 3000-4000 tons leading an FTD estimated at 1500-2000 tons towing a Type X fuel barge. These lay about one mile behind the FTC, with two escort vessels weaving within them. The escort vessels appeared to be about 150 feet long and were either patrol craft or submarine chasers.

"Descending to 6,000 feet to avoid cloud, the Liberators then proceeded to attack in line astern formation. At 020240Z, No 3 duty was assigned to the lead position with the task of dropping one bomb to commit the FTC to evasive action and thus define a new course upon which the following duties could bomb. However, the bomb dropped by Duty No 3 struck the vessel aft of the stack and caused large volumes of black smoke to appear immediately. The ship was seen to roll over on its side and sink within 30 seconds.

"All five duties then commenced individual attacks on the FTD and the fuel barge. The FTD cast the barge adrift and proceeded to take evasive action. One escort vessel steamed off to a distance of about 3 miles, whilst the second vessel taking violent evasive action remained in the vicinity of the shipping under attack.

"All ships maintained ineffective light to medium ack-ack; shells were seen exploding at about 4,000 feet and also falling into the water. The barge appeared to be firing a machine gun from the bow. A number of near misses and two probable hits were scored on the FTD which was left stationary, down at the stern and apparently in a sinking condition. Two Liberators reported seeing lifeboats lowered by the vessel.

"Three bombs were dropped towards the fuel barge resulting in two very near misses and a possible hit on the bow. It was left down by the head with a large oil slick in the water. All bomb loads were expended by 020325Z, and as the Liberators departed the area, they observed the escort vessels moving in towards the disabled vessels.

"Tasks were completed, and the Liberators landed at Truscott at 020825Z".

One of the Liberators taking part in the mission described above came to a sticky end exactly two months later. On 2 February 1945, A72-88, in company with A72-54, 73, 77, 84 and 93, were ordered to execute a shipping strike against a 7000 ton enemy vessel which had been reported in the Flores Island area of the Dutch East Indies.

Flying from Fenton, the Liberators failed to find their intended victim but instead came across a 700 ton vessel which was attacked and left on fire. Four of the aircraft left the area with the intention of flying to Truscott, while the other two (including A72-88) returned to Fenton, a considerable distance to the south-east.

A72-88 arrived over Fenton in hazy conditions with limited visibility and on the downwind leg of its landing circuit could not see the runway flare path. On finals, the flarepath was visible only from below 700 feet. The squadron report continues the story: "The captain indicated that another circuit would be made and ordered the retraction of the undercarriage. Without warning the aircraft swung violently to port and the captain endeavoured to correct this.

"He then noticed that the fuel pressure on Nos 1, 2 and 3 motors had dropped to zero and that fuel pressure on No 4 motor was fluctuating. Power was increased on

No 4 motor in an endeavour to maintain height.

"A72-88 lost height rapidly and could not be kept on a straight course. At 200 feet the captain switched to intercom and warned the crew that a crash landing was imminent. The aircraft crashed through trees and landed on a level keel approximately 2 miles south-west of Fenton where it caught fire".

21 Squadron RAAF

Following a 12 month period operating Vultee Vengeance dive bombers, including active service in New Guinea, No 21 Squadron was recalled to Australia in mid 1944 to be reformed as a heavy bomber squadron equipped with Liberators. Initially headquartered at Camden NSW and then Leyburn Qld, the squadron began receiving Liberators in September 1944 while under the command of Grp Capt P A Parker.

Training requirements and a shortage of aircraft delayed 21 Squadron's operational debut on the Liberator until January 1945 by which time the unit was part of the RAAF's No 82 Wing and based at Fenton in company with No 24 Squadron. The first Liberator operation was carried out on 11 January 1945 – an armed reconnaissance around the Dutch islands to the west of Timor which lasted 12 hours and resulted in no sightings of enemy shipping or activity.

From then, 21 Squadron's activities were similar to those of 24 Squadron and 82 Wing, including the sending of a detachment to Morotai Island in April. From there, the squadron covered allied landings in Borneo and 'softening up' operations there and Labuan and Balikpapan in preparation for the invasions of those areas. The part of the squadron which had remained in the Northern Territory continued operations from there. The Morotai detachment bombed targets in the Halmaheras and Celebes Islands during the last weeks of hostilities and on 12 August carried out its final operation of the war, a raid on Lake Lina, Halmahera.

With the end of the fighting, 21 Squadron indulged mainly in transport duties including the repatriation of Australian prisoners of war to Australia. The squadron moved from Morotai to Tocumwal NSW in November 1945 from where transport duties continued, albeit with a much reduced personnel strength and only four Liberators by February 1946.

In April 1946, No 21 Squadron moved to Amberley Qld where it remained until January 1948 when it was disbanded and its men and machines posted to reform No 2 Squadron equipped with Lincolns. The squadron's time at Amberley was punctuated with uncertainty as strength was small with only two crews on the books. This temporarily improved later in 1947 when a VIP Flight was established which undertook journeys to Japan, New Zealand, New Guinea and the Philippines.

23 Squadron RAAF

The third of the RAAF squadrons which made up No 82 Wing during 1945, No 23 was another unit which had operated Vengeances in New Guinea before becoming a heavy bomber squadron equipped with Liberators.

It, too, was withdrawn to Australia during mid 1944, initially to Menangle Qld and then in October to Leyburn where re-equipment with Liberator gradually took place under the command of Wng Cdr R A Dunne, who remained CO until the war had ended. But there were delays, so it wasn't until April 1945 that the squadron was operational and flying from Long, near Fenton in the Northern Territory. The squadron then moved to Darwin in July and a detachment joined those from Nos 24 and 21 Squadrons on Morotai.

After fighting had ceased, 23 Squadron was transferred to Tocumwal NSW in November 1945 and then to Amberley Qld where it was re-equipped with Lincolns in mid 1947 and renamed No 6 (Bomber) Squadron in February 1948.

Despite being a relatively 'late starter' in the Pacific war, No 23 Squadron made a substantial contribution to the Borneo and Balikpapan campaigns in conjunction with other RAAF and USAAF Liberator units. These activities will be discussed later in this chapter.

12 Squadron RAAF

The next RAAF unit to receive Liberators was No 12 Squadron, another which had previously operated Vengeances in the New Guinea area. After being reduced to a cadre unit in the interim, the squadron moved to Cecil Plains near Dalby Qld in December 1944 and began receiving its first Liberators. It remained there until re-establishing at Darwin in May 1945 from where operations began. In addition, a detachment was further south at Fenton in company with the Liberators from 82 Wing's Nos 24 and 21 Squadrons. The commanding officer at the time was Wng Cdr N G Hemsworth.

Operations from Darwin were similar to those which had been conducted by the other Liberator squadrons and mainly comprised attacks against Japanese shipping and outposts in the Timor, Arafura and Banda Seas areas.

The end of hostilities again repeated the experience of the other Liberator squadrons with transport duties (including the dropping of supplies to POWs all over the Dutch East Indies and their subsequent repatriation) occupying most of the flying hours.

No 12 Squadron (along with No 99) became part of the RAAF's newly formed No 85 Wing in July 1945.

The squadron moved back to Australia in February and March 1946, establishing itself at Amberley and like the other Liberator squadrons which joined it there (Nos 21 and 23) it was much reduced in both its personnel and aircraft strengths. Conversion to Avro Lincolns began in June 1947 and the new type operated alongside the remaining Liberators until the beginning of 1948 when the Lincolns took over completely. In February, No 12 Squadron ceased to exist as such, its personnel and equipment forming the basis of the reformed No 1 Squadron.

One of the last flights performed by a 12 Squadron Liberator was a special one. In December 1947 one of the aircraft was used to fly the Australian Prime Minister, Mr Ben Chifley, to New Zealand. The captain of the aircraft was Wng Cdr John Hampshire DFC, a junior officer with the squadron at Darwin in 1939/40 (whilst flying Ansons) and later the commanding officer of No 25 Squadron and its Liberators.

99 Squadron RAAF

The second flying element of 85 Wing, No 99 Squadron was newly formed in February 1945 at Leyburn Qld, but became operational too late to see any combat. It was originally intended that 99 Squadron would be the RAAF's first operational Liberator unit, but revised plans saw that come to nought.

The squadron's first Liberator (A72-168) arrived in March 1945, the same month training began and the squadron moved to Jondaryan, pending a more permanent shift to Darwin.

By mid April 99 Squadron had eight Liberators and 18 complete crews on strength and was under the command of Wng Cdr A E Cross. The squadron began its move to Darwin in May but the process was somewhat leisurely, with it not

The crew of A72-81 on the Kali Konto power station strikes. Standing, left to right: Sgt M W Lewis, Flt Lt G A Bond, Sqdn Ldr S J Nichol, Flg Off E T Jones, Flt Lt P Tanner, Sqdn Ldr H H Stelle, Wnt Off J Holliday, Grp Capt D W Kingwell, Flt Sgt R Hogan. Kneeling: Wnt Off R L Baker. (via Neil Mackenzie)

becoming fully established there until the middle of September, by which time hostilities had been over for a month and Japan's surrender had been signed.

The squadron's major activities revolved around transport duties for the next few weeks and included the carriage of 495 prisoners of war, 299 other passengers and 42,614lb (19.3 tonnes) of freight from Darwin to Australia's major cities. Return trips to Darwin carried twice that amount of freight.

No 99 Squadron moved to Tocumwal in November 1945, preparatory to its winding down and eventual disbanding in June 1946.

25 Squadron RAAF

The fifth and final former Vultee Vengeance squadron to convert to Liberators, No 25 spent all of the war years based in Western Australia, mainly at Pearce near Perth. The switch to Liberators was made in early 1945 when the squadron was under the command of Wng Cdr N G Hemsworth, and at the same time a move was made to Cunderdin, about 70 miles (110km) inland from Perth and not far from Pearce.

25 Squadron's main roles during the war had been convoy support and anti submarine patrol sorties. With the arrival of Liberators, the emphasis changed to bombing raids on targets in the Dutch East Indies, staging through airfields in the Northern Territory. Many of these raids were conducted against Japanese airfields from which attacks on Morotai Island were being launched, while 25 Squadron's first offensive mission with Liberators – on 13 March 1945 – was conducted by six squadron aircraft in conjunction with a similar number from 82 Wing against Mapin, a barge staging point on Sumbawa Island.

24 Squadron's LAC Bannah, an armourer, cleans the ball turret guns of A72-40 between missions. (via Neil Mackenzie)

Maritime activities continued, however, including an incident in February 1945 involving the sinking of the motor vessel *Peter Sylvester* by the German submarine U.862 which had been achieving some success against convoys in the Indian Ocean off the coast of Western Australia.

U.862 torpedoed and sunk the *Peter Sylvester* on 6 February some 800 miles (1,300km) off Fremantle and although 15 survivors were picked up by another ship three days later, others were still unaccounted for.

Some of No 25 Squadron's newly delivered Liberators were allocated to the search, and on 10 February rafts with 20 survivors aboard were spotted by one of the Liberators which dropped rations and an emergency radio transmitter. The Liberator returned to base after a sortie lasting over 16 hours. Three days later, another 25 Squadron Liberator spotted more rafts and a lifeboat containing between them 30 survivors. After dropping supplies, the aircraft guided an American vessel to the scene and they were picked up.

One hundred and forty three of the *Peter Sylvester's* complement of about 170 was eventually rescued, although 15 Squadron suffered its own tragedy during the course of the operation when A72-124 crashed and burned whilst taking off. Five of the crew was killed. The aircraft settled back onto the runway just after takeoff from a height of about 30 feet. It bumped three times, skidded through a fence and exploded. The Liberator was fully loaded with fuel for its lengthy mission, including a 700 gallon (3,180 l) auxiliary tank in the bomb bay.

25 Squadron's efforts against targets in Java and elsewhere in the Dutch East Indies were often conducted with aircraft from No 23 Squadron, while those from Nos 24 and 21 Squadrons at Morotai concentrated on Borneo.

25 Squadron lost its first aircraft over enemy territory on 26 April during an eight aircraft raid whilst attacking Malang airfield near Surabaya. The raid was led by Sqdn Ldr J E S Dennett, the temporary commanding officer (Wng Cdr John Hampshire took over in May) who had to cope with a heavy cloud cover over the target. Sqdn Ldr J A Wawn's Liberator (A72-133) sent out distress signals and had to make a forced landing, whereupon the crew was captured. Despite the usual brutal treatment of the men by the Japanese (particularly Wawn, who was tortured), all survived to be freed at the end of the war.

In common with the other RAAF Liberator squadrons, No 25's aircraft were extensively used for transport and POW repatriation duties when the fighting stopped and the squadron survived intact only until March 1946 when it ceased flying. It was formally disbanded the following July.

Maintenance work on 24 Squadron Liberators at Fenton, with just about every hatch open. A72-38 was a B-24J delivered in June 1944 and scrapped in 1952. (via Neil Mackenzie)

102 Squadron RAAF

102 Squadron was another RAAF Liberator unit formed too late to participate in the war against Japan. Formed at Cecil Downs Qld in late May 1945, the squadron didn't receive its first Liberator until July and its active service comprised transport duties between Amberley and Finschhafen, Morotai and Bougainville during the last four months of 1945.

With no further need for its services, the squadron was wound down in the early part of 1946 and disbanded in March.

82 Wing RAAF

The RAAF's main heavy bomber Wing, No 82, was formed at Ballarat Vic in August 1944 under the command of Grp Capt D W Kingwell with the intention of taking over the duties of the USAAF's 380th Bomber Group which would be moving from the Northern Territory to the Philippines where it would join other elements of the 5th Air Force.

The wing's operational headquarters was established at Fenton NT in October 1944 and operations began the following January. At that stage 82 Wing comprised Nos 24 and 21 Liberator squadrons, No 24 Air Stores Park, No 27 Medical Clearing Station and No 18 Repair and Salvage Unit.

During the course of 1945 it was decided to move 82 Wing to Balikpapan on Borneo once it had been wrested from Japanese control. 82 Wing's Liberator squadrons (by now joined by a detachment from No 23 Squadron) began operating from Morotai in June in support of operations against Borneo, conducting numerous raids particularly in and around Balikpapan preparatory to invasion in company with large numbers of other Allied aircraft. By now, 82 Wing was part of the First Tactical Air Force.

As it happened, 82 Wing remained based on Morotai until the end of the war, the Liberators never moving to Balikpapan despite advance parties going there in July 1945.

Postwar, 82 Wing was temporarily established at Tocumwal before moving to Amberley where it operated with Nos 12, 21 and 23 Squadrons under its command. By early 1948 the Liberator squadrons had been disbanded and the wing would reform with Nos 1, 2 and 6 Squadrons, all equipped with Lincolns and established from the equipment and personnel of the former Liberator units.

"Engine attention for mighty RAAF Liberator ..." says the official caption to this photo of a 21 Squadron aircraft undergoing some minor servicing. In the background is A72-53, a B-24J delivered in August 1944. (via Neil Mackenzie)

Nos 200 and 201 Special Duties Flights

An interesting aspect of any conflict is the covert operations which are carried out in all theatres of war. The South-West Pacific Area was no exception, the need to gather information and to insert, retrieve and supply those involved in such activities was great. It began with the coast watchers, that extraordinarily brave group of people who kept tabs on Japanese movements at great risk to themselves, and developed that to include those involved in guerilla activities as well as just collecting information.

For most of the Pacific war, airborne activities along those lines mostly fell to the British, although Australian Hudsons and Catalinas drawn from regular squadrons and flights were also involved. It wasn't until early 1945 that a special duties flight within the RAAF was established to provide such a service, and the Liberator was intimately involved.

Following some success with paradropping personnel behind enemy lines from a modified 24 Squadron Liberator in the second half of 1944, it was decided by the Allied Intelligence Bureau that it should have its own flight of aircraft. After some months, approval was given in early 1945 to form No 200 (Special Duties) Flight equipped with six Liberators and initially based at Leyburn Qld. The flight came into existence on 15 February 1945.

200 Flight's first commanding officer was Sqdn Ldr Pockley DFC (a former Sunderland pilot with 10 Squadron in Britain) and its first Liberator was A72-191. By mid March the flight had its full complement of six Liberators on strength, all of which incorporated modifications

No 21 Squadron was the RAAF's second operational Liberator unit, coming on strength in January 1945 and joining No 24 Squadron as part of 82 Wing. Following a raid on Labuan Island, this Liberator is having its spent shells removed from the nose turret while ammunition for the next sortie is brought in for loading. (via Neil Mackenzie)

No 23 Squadron made up the third flying element of 82 Wing but didn't become operational until April 1945. Three of its Liberators are photographed over Darwin Harbour in July 1945. (Bruce Chapman)

No 23 Squadron B-24M Liberator A72-167 over the Northern Territory coast. The 'V' on the vertical tail is a formation stripe. Some RAAF Liberators carried them, each squadron having the 'V' in a different position – inverted (99 Squadron), base of 'V' forward (24 Squadron) and base of 'V' to rear (21 Squadron). (Bruce Chapman)

such the fitting of jump lights and a parachutist's exit chute. Unlike the special Liberators flown by Royal Air Force units in the area, 200 Flight's aircraft carried their normal full crew of 11 and retained their usual defensive armament although the ventral ball turret was removed after a time.

An early operation sums up much of 200 Flight's activities in the final months of the war: On 15 March 1945, three Liberators were despatched on a mission to drop agents and stores into Borneo. The mission was a success despite two aborted attempts due to poor weather and visibility. On the third attempt, the personnel and equipment were successfully dropped from relatively low level over terrain which was anything but hospitable and in weather which could close in without warning.

A 'hung up' load caused some problems for one of the aircraft but eventually all the personnel and all but one load of stores landed in the small clearing which had been selected as the drop zone. Unfortunately, the success of the mission was marred by the loss of Pockley and his crew in A72-191 on the return journey, and an extensive search failed to find any trace of the Liberator or the 12 people on board. Two other Liberators failed to return from missions whilst in the service of 200 Flight, A72-159 and A72-177, both in May 1945.

The flight developed its expertise and increased its sortie rate from April 1945, now under the command of Wng Cdr Read. For dropping, the usual technique was to fly at 135mph with half flap and reduced power on the inboard engines. Dropping height could vary from 300 to 3,000 feet above ground level.

A problem was the distance the flight had to travel from its Queensland base to its area of operation mainly in the Borneo area. Repeated requests to move to a more forward base such as Morotai or Noemfoor were denied (although some missions were flown to Morotai) and 200 Flight remained at Leyburn until it was disbanded on the last day of 1945.

Liberators known to have flown with No 200 Flight are A72-159, 177, 180, 182, 183, 185, 187, 191 and 192.

Another little known unit which flew Liberators was No 201 Flight, formed at Laverton, Victoria on 10 March 1945, under the command of Wing Commander C S Davis DFC. The initial aircraft allocation was two B-24Ms, A72-186 and -197. These aircraft were handed over to 1 Air Performance Unit for modification, which included the removal of the ball turret, the installation of a radar cubicle and the fitting of AN/APR-4 receivers. Delays in modification were experienced in securing the necessary intercept equipment. It could be said that these aircraft were forerunners of today's ELINT (electronic intelligence gathering) types.

On 17 April 1945, a detachment of eight officers and 80 airmen departed for Darwin, arriving at Winnellie on 30 April. The detachment remained at Darwin until they were withdrawn after the cessation of hostilities.

A72-357 was allotted to 201 Flight to enable crews to train whilst the other two aircraft were being modified. After modifications were completed, the unit flew flight trials out of Laverton, before ceasing to function on 17 December 1945. The unit was disbanded on 15 March 1946.

The *Isuzu* Sinking

Mention was made in the Mitchell section of this book of a mission on 6 April 1945 against the 5,700 ton Japanese light cruiser *Isuzu* in the Flores Sea. The ship – accompanied by four smaller vessels – was spotted steaming towards Koepang on Timor where it was assumed a short stop would be made to pick up evacuees. The convoy was photo-

Liberators from two of the squadrons which made up 82 Wing – Nos 24 (GR code) and 21 (MJ) photographed en route to Labuan Island off North Borneo. RAAF Liberators played a major role in the Borneo campaign. (via Neil Mackenzie)

graphed by Mosquitoes of the RAAF's No 87 PR Flight and then shadowed by Catalinas from No 43 Squadron before being attacked by twenty B-25 Mitchells (ten each from No 18 Squadron NEIAF and No 2 Squadron RAAF – the components of No 79 Wing) and nine Liberators from 82 Wing. Of these, five came from No 24 Squadron and four from No 21.

The attack was made in the early hours of the morning just after the *Isuzu* left Koepang, 18 Squadron's Mitchells achieving two direct hits but the Liberators and Mitchells from No 2 Squadron failing to repeat that success. The damaged cruiser temporarily escaped, and was sunk by an American submarine the following day.

While the Mitchells experienced inaccurate anti-aircraft fire and survived the attentions of two Japanese fighters, by the time the Liberators arrived 20 minutes later, the ack-ack had found its range and the fighters were more determined.

The RAAF Official History takes up the story: "... the nine Australian Liberators, led by Group Captain Parker, attacked from 12,000 feet. This time the anti-aircraft fire was extremely accurate and bursts were seen all around the third element (Sqdn Ldr G J White, Flt Lt S L McDonald and Flt Lt Court) during its first run over the target at 10am. As it did so, enemy fighters attacked the formation, one closing to within 20 feet of Court's aircraft, in spite of which the Australian gunners did not appear to have scored any hits on it.

"Two fighters attacked McDonald's

Liberators of No 25 Squadron at Cunderdin WA in October 1945. The black swan motifs on the aircrafts' tails are appropriate because the squadron spent the war exclusively based in Western Australia. The aircraft in the foreground is B-24L A72-150, delivered in December 1944. (via Neil Mackenzie)

aircraft, closing to within 50 feet and breaking away. Cannon shells burst under the flight deck of the Liberator, near the nose wheel. Fire broke out immediately and spread very quickly. McDonald then ordered the crew to bale out.

"During the second run over the target, the enemy fighters attacked again and Flt Lt Ford's aircraft was hit. A Japanese fighter also appeared to be hit. One of the Liberator's motors caught fire and flames were coming from the nose wheel compartment. Flames were licking around the feet of Warrant Officer C G Vickers, the second pilot, when Ford gave the order to abandon the aircraft. The back hatch was jettisoned and five of the crew managed to jump as the aircraft was held on a level course by Ford. However, the Liberator then rolled over and, with Ford still at the controls, plunged into the sea, exploding on impact".

A further view of 200 Flight's A72-183 undergoing maintenance. The flight operated a total of nine Liberators of which three failed to return from missions.

Photographs of 200 and especially 201 Flight aircraft are not common, but both are illustrated here. B-24M A72-183 (above) of 200 Flight undergoes some servicing while B-24J A72-357 (below) of 201 Flight waits at Laverton. (David Wilson/Neil Mackenzie)

Remarkable photograph of the downing of one of the two Liberators lost during the April 1945 attack on the Japanese light cruiser 'Isuzu' in the Flores Sea. Intense anti aircraft fire and fighter protection were common hazards in attacks on Japanese shipping. (RAAF)

Warrant Officer K R Shilling, on board Flt Lt McDonald's doomed Liberator, recounted the drama and his own considerable ordeal:

"The fire drove the engineer into the bomb bays. I did not see him again. The captain ordered the crew to bale out. The second pilot escaped through the bomb bays. I followed 30 seconds later. About two minutes later another member jumped. He did not appear to have a parachute. I went out of the bomb bay head first, counted six and pulled the release cord. It took me about 12 minutes to come down, during which time I counted six members parachuting from Flt Lt Ford's Liberator. An additional member appeared to drop without a parachute.

"McDonald's aircraft disintegrated about 3,000 feet above the sea. It did not explode. Ford's Liberator exploded on impact with the water. Enemy fighters did not attack the parachutists. My legs were entangled in the shroud lines and I had to cut away the lines with a Bowie knife. Although I had seen someone land about 50 yards away heavy seas prevented my seeing him. The seas were running at about four and a half feet.

"I never saw any member of my crew again.

"After about 15 to 30 minutes the enemy cruiser came within 100-200 yards of me. She was heavily laden with troops. Equipment under tarpaulins was clearly visible on the decks. A machine gun opened fire on the position I was making for. I deflated my Mae West and sunk. At this time the bullets hit the water within six feet of me. The cruiser then passed on. Some 15 minutes later an escort vessel of about 1,000-2,000 tons passed within a hundred yards of me. This vessel was also laden with troops and equipment. It continued on for five minutes, turned, and came directly towards me, passing close enough for the bow wave to throw me sideways. I could hear voices from the deck. No attempt was made to pick me up or shoot me.

"Some minutes later a Zero passed overhead low and appeared to be looking for survivors. I deflated my Mae West again and sank to avoid detection. About ten minutes later seven Liberators circled low and after some five minutes dropped me a dinghy which landed 50 yards away. Twenty minutes later a Catalina [from No 112 Air Sea Rescue Flight] arrived and taxied close enough to pick me up ..."

Already on board the Catalina (Flt Lt R Bulman, captain) were Sgt W W T Sayer from Flt Lt Ford's crew and Wnt Off Vickers from the same crew was also picked up shortly afterwards, as was Flt Sgt Faichnie.

At this point it appeared that everything might be rosy for the survivors, but more was to come. Wnt Off Shilling continues:

As the completely exhausted Faichnie "... was being pulled in over the gun blisters the Catalina was attacked by a Zero from 12 o'clock. The Catalina caught fire immediately and sank three minutes later. I had undressed, as ordered, and going to bed in a bunk in the waist. I had been called forward for the landing and was in the navigator's position when the attack was made. The order was given to bale out and I made my way aft. On the way a fuel line burst above me, pouring blazing fuel on to my back. This was extinguished by Sgt Sayer who smothered the flames with a sleeping bag.

"I escaped out of the port blister, naked and with no Mae West. I do not remember much for a few minutes and when I regained my facilities I was being kept afloat by two members of the Catalina, Flt Lt Bulman and Flt Sgt Scholes. These two men had to let me go after 15 minutes owing to their Mae Wests coming undone. Flg Off Becke (another of the Catalina's crew) helped for the next 45 minutes. Through the fortitude of these men I was saved from drowning ..."

Shortly after the Catalina was hit, a Liberator dropped a dinghy and supplies to those in the water and after a lengthy battle with the heavy seas and wind, another Catalina arrived to pick up the three survivors from the Liberators and the six members of the first Catalina. As incredible as it may sound, this Catalina was also attacked by a Japanese aircraft, this time an Nakajima J1N1 'Irving' twin engined fighter-bomber, just as the Catalina was taking off. The Catalina received some hits but survived them and the running fight which followed. Eventually, the Catalina made it back to Darwin.

An amazing story.

Pushing Them Back

Although the RAAF's Liberator squadrons conducted countless sorties all over their area of operations, several campaigns during 1945's 'pushing back' of Japanese forces and reclaiming territories from them deserve special mention and give some idea of the important role these aircraft played in the last year of the war.

The island of Tarakan off north-east Borneo was a vital step in the Borneo campaign as it would provide a base for subsequent operations against Balikpapan and for convoy protection in the area. Additionally, it was major centre of oil production and processing capable of turning out 6,000,000 barrels a year.

Among the numerous air elements employed against Tarakan preliminary to a landing by troops were the Liberators of Nos 21 and 24 Squadrons, RAAF. Their job was to attack personnel emplacements including barracks plus other buildings and equipment. Although it

Two shots of RAAF Liberator action against Japanese shipping. A lugger (above) is bombed and strafed by a 12 Squadron Liberator in the Flores Sea during June 1945. The photograph was taken from a height of 200 feet. The shot (below) was taken from a 21 Squadron Liberator and shows an attack on a 'Sugar Dog' light freighter in the Flores Sea in February 1945. Both photographs illustrate attempts at camouflaging by the Japanese. (RAAF Historical)

(right) 21 Squadron's Sqdn Ldr G J White (right) in front of his Liberator with symbols of raids and successes painted on the nose of the aircraft. White was one of those involved in the 'Isuzu' raid. (L E Benoit via Mike Kerr)

was inevitable that some of the oil producing facilities would be damaged, it was intended that they emerge from it all as unscathed as possible, as the oil was as useful to the Allies as it was to the Japanese.

The first raid by 82 Wing's Liberators was flown from Morotai on 13 April 1945 and by the first week of May the airfield was in friendly hands, although it then took rather longer than anticipated to bring it to a serviceable state.

The next step was Labuan/Brunei Bay, on the north-western coast of Borneo, although in the time leading up to that assault, 82 Wing's Liberators concentrated their efforts on the Balikpapan/Samarinda area on the east coast, attacking gun positions, oil tanks, buildings and airfields. It was all part of the 'softening up' process which would soon see Borneo in Allied hands.

On the day of assault landing on Labuan – 10 June – all three of the 82 Wing's Liberator squadrons (Nos 21, 23 and 24) were involved in high level bombing of the area immediately behind the beach head with anti personnel bombs. By the end of the day the battle for Labuan had been won.

Balikpapan

Three weeks after the taking of Labuan, Balikpapan became the next Borneo target for invasion. Once again, oil was the major resource and the position of Balikpapan meant that the stronghold on the

The first bomber to land at Jacquinot Bay, New Britain, on 9 April 1945. B-24M A72-185 was carrying Senator J M Fraser, the acting Minister of Defence. This Liberator also served with 200 Flight. (via Neil Mackenzie)

An RAAF Liberator undergoing maintenance at Palawan in the western Philippines during the Borneo campaign. A USAAF B-25 Mitchell and P-38 Lightning can be seen in the background. (via Neil Mackenzie)

They also served. Often forgotten is the contribution Australian airmen made to the India-Burma campaign, hundreds of them serving with Royal Air Force squadrons in the area. Of those, many served on RAF Liberators, such as this one from No 356 Squadron, photographed after bombing a Japanese oil refinery. (via Neil Mackenzie)

The Australian influence on this RAF Liberator operating in South-East Asia is obvious with the name 'Black Swan' adorning the aircraft's nose. (via Neil Mackenzie)

The Liberator began its career as a transport and that's the way it ended for many of them postwar. Qantas used Liberators on its Indian Ocean services for ten months from July 1945, replacing Catalinas and complementing Lancastrians. The aircraft were used on the Sydney/ Singapore run for a time after that. VH-EAI was the former RAF Liberator LB-30 AL541.

The end of the road. Liberator 'graveyards' at Tocumwal NSW (above) and Amberley Qld (below) in the early 1950s. More than 200 of the RAAF's 287 Liberators were turned into aluminium ingots and others were used for fire training. (J de Horn via Mike Kerr)

Japanese in the area would be further tightened if it were taken. Balikpapan had been the area's second largest oil production centre before the war.

The occupation of Balikpapan was a major operation involving a large number of Allied aircraft of all types and an armada of 200 American, Australian and Dutch ships plus thousands of troops who would make an amphibious assault.

Balikpapan had received considerable attention from American and Australian bombers over the previous few months but the oil producing facilities were still operating and resistance had not crumbled. The invasion was scheduled to take place on 1 July 1945 and in the two weeks before that, the place would have to endure a massive bombardment from the air and from the sea.

All three RAAF Liberator squadrons from No 82 Wing were involved, and it's interesting to note the general orders for the Balikpapan operation included the words "... seize and occupy the Balikpapan, Borneo, area, establish naval, air and logistic facilities and conserve petroleum producing and processing installations therein ..."

Flying from Morotai, over 700 miles (1,100km) away, the Australian Liberators conducted numerous raids on anti aircraft positions, coastal defences and some of the oil installations. On invasion day Australian and American Liberators bombed the beach defences, the 20 RAAF Liberators (seven from Nos 21 and 24 Squadrons plus six from No 23) dropped 500lb bombs on targets near the beaches. The mission was led by Sqdn Ldr I O Black.

The Australians' contribution to what amounted to the decimation of Balikpapan was a tiny one. Subject to incessant bombing and a mighty naval barrage, the town was just about wiped off the face of the earth, and the oil refinery and associated facilities lay in ruin ... not quite what had been ordered. The devastation was so complete the Dutch (whose property it had been before Japan invaded) complained that the bombardment of Balikpapan was – to use a modern expression – "way over the top" and the damage it caused unnecessarily great.

If nothing else, it did ensure a relatively safe landing for the troops.

★ ★ ★

Of the 287 Liberators delivered to the RAAF in 1944 and 1945, only one – or at least its fuselage – survives. A72-176, a B-24M flown by the former CO of 82 Wing, Air Commodore Deryck Kingwell CBE DSO, while he was Officer Commanding East Sale and Tocumwal in 1946-47 (and known as "Kingwell's Kite"), was delivered to the RAAF in January 1945 and after being struck off charge in March 1948 was transferred to RAAF East Sale as an instructional airframe. In the same year, its fuselage was offered for sale as scrap and purchased by Mr George Toye of Moe, Victoria.

Mr Toye has owned the Liberator ever since, kept under cover and well preserved on his property. The aircraft is now subject to a restoration programme, using a set of wings found in Alaska. When completed, the aircraft will be displayed in an accredited aviation museum as a suitable memorial to the RAAF's Liberators and the men who flew them.

One that got away. Deryck Kingwell and owner George Toye sit in the cockpit of A72-176, the fuselage of which has been kept at Mr Toye's property since 1948. This particular Liberator was flown by Kingwell in 1946-47. (RAAF Historical)

CRASH CRITIQUE
DITCHING OF B-24J LIBERATOR A72-348

This "crash critique" deals with the ditching of A72-348 due to engine failure during a travel flight from the Philippines to Darwin. Its purpose was to pass on a number of important points which were learned and which, if studied, would prove of value in the case of another similar incident. Squadron commanders were recommended to read the critique to their pilots and use it as a basis of discussion. The critique also highlights the search and rescue effort which was mounted. Some parts have been omitted.

Aircraft: B-24J Liberator A72-348
Captain: Wing Commander J E Handbury
Complement: 8 crew and 2 passengers
Date: 9 April 1946
Aircraft takeoff weight: 59,000lb
Nature of flight: Travel flight from Nichols Field Manila to Darwin
Fuel: Full tanks

The aircraft took off from Manila at 0650I for Darwin. Until the time of the accident the engine performance was normal. In flight, wireless contact was established with Morotai and weather conditions were forecast as favourable.

Power used for takeoff was 48in/2700rpm, and the aircraft then climbed to 9,000 feet on climbing rpm. At this height, cruising power of 32in/2200rpm was used for the first two hours. Mixture Automatic Lean was used after approximately 30 minutes in this combination. After this period, power was further reduced to 32in/2100rpm, and this power setting remained unaltered until the time of the engine failure.

At 1040I, without any warning or any apparent unusual engine performance, No 5 (top) cylinder of No 2 engine exploded, carrying away the upper cowlings. Immediately, the standard procedure was adopted and an attempt was made to feather the airscrew but this operation could not be done. Almost simultaneously, No 1 engine lost power, the cause of which is unknown, followed by excessive vibration of the port mainplane. Feathering was then attempted on No 1 engine, but again the feathering system would not operate.

Combined with the inability to feather, the revs on engines 3 and 4 could not be increased or decreased. However, the decision made for an immediate ditching was directly contributable to the excessive vibration of the port engines and mainplane. Wing Commander Handbury informed the crew at 8,000 feet that a ditching may be necessary and they were instructed to dispose of all loose gear by dumping. At a lower altitude, a further order was given to take "ditching stations".

The navigator was instructed to obtain a position, which was 0500N, 12457E. The wireless operator despatched this information as an emergency distress signal. This was acknowledged by Morotai.

The following is an account of the actual ditching and conditions existing at the time:

(a) Weather conditions: good, with wind of approximately 8 knots from the east. The sea was calm and a ditching was made into the swell with a 70deg crosswing.

(b) Method of approach and ditching: height was lost from 9,000 feet to the water without turning owing to the difficulty in controlling the aircraft. The final approach speed was 150mph, with full flap and limited power on the remaining serviceable engines. This speed was used to give the pilot a longer period to estimate the height above the water.

The attitude of the aircraft for ditching was slightly nose up position. The effect of the ditching was a planing of the rear portion of the fuselage with an almost immediate pitch forward causing the nose of the aircraft to break up. At the same time, the fuselage aft of the command deck broke cleanly with the nose of the aircraft under the water.

Ditching stations: Captain and 2nd pilot – normal position and securely strapped in; Flight Engineer – braced rear of captain for the purpose of operating automatic dinghy release.

It now appears that the wireless operator and navigator continued their efforts in despatching emergency signal and position report too long which caused them to remain in their normal flight position. Remainder of crew and passengers were aft and exact positions of all is not known but at least four were braced opposite side windows against a platform which was put over the ball turret position for this trip.

In the ditching, the navigator who was on the flight deck and one of the passengers aft of the command deck lost their lives. Other members sustained cuts and bruises, but the wireless operator was severely injured in his position by the wireless table. The slight injuries are thought to have been caused by: members not bracing themselves sufficiently for the impact; taking station at other than a recognised ditching station and consequently being thrown forward; relaxing on the aircraft first striking the water in a planing attitude, not realising that a major impact was to follow.

Immediately after ditching, the flight engineer pulled the dinghy release behind the pilot on the flight deck, and members of the crew in the front of the aircraft escaped through the broken top portion of the cockpit, whilst other occupants escaped through a break in the fuselage aft of the command deck.

Members of the crew aft recovered the Gibson Girl and thermos flask from the aircraft. On escaping from the aircraft, it was found that the dinghies had not released automatically so two members of the crew manually released them, even then, one dinghy only half inflated and had to be fully inflated by hand.

The injured were then helped into the dinghies and a thorough search was made for the missing, of whom no trace was found. On completion of the search, the two dinghies were tied together with four occupants in each. The Gibson Girl was then assembled and was operating within an hour of ditching. The delay in assembling the Gibson Girl was due to the hydrogen generators being damaged and difficulty was experienced in inflating the balloon.

During this time, first aid was given to the injured. It was noticed that the bandages from the two medical kits were not sufficient.

Normal dinghy procedure was carried out and duties allocated to slightly injured members, eg: operation of Gibson Girl, watch for search aircraft and/or ships, attending the seriously injured, stand by for pyrotechnics. A fishing line was also put out and a sea anchor streamed.

The operation of the Gibson Girl was discontinued at approximately 2200I hours, on 9 April, until a flare fired by HMS *Opussum* was sighted at approximately 0230I. Immediately on sighting this flare, the crew returned a red parachute flare. It is now known that the first flare was not seen by the *Opossum* but the second answering flare fired some time later was. The Gibson Girl was then operated for visual light signals on which the *Opossum* homed. The survivors were rescued at 0330I on 10 April; time of rescue from when the *Opossum* saw the flare approximately 1 hour.

Remarks and Recommendations of Aircraft Commander

Ditching positions: Captain and 2nd pilot in normal position; flight engineer braced behind captain clear of mid-upper turret; wireless operator and navigator to go aft and brace on command deck facing aft with rest of crew.

Wireless operator and navigator are not to be sent to this position unless there is ample time to enable them to take position before the aircraft ditches. When passengers are carried, ditching positions should be allotted by the captain where solid bracing aft of the command deck is available, and further, captains should stress the violence of the final impact.

Whilst correct ditching and dinghy drill should always be carried out when conditions are normal, all personnel should be prepared to adapt themselves to unusual circumstances and all members should check that the dinghy, Gibson Girl etc, are being looked after by those detailed.

When an aircraft passes an emergency distress signal with a position or the base shows that an aircraft has ditched in a limited area, it is considered that a search at night, both by ship and by aircraft, provided weather conditions are favourable, is not impractical and for the location of dinghies, preferable.

For instance, in a night search, a ship could fire a rocket or snowflake every 30 minutes, and an aircraft a two star cartridge say every five minutes. From the dinghy, the snowflake fired from the *Opposum* was seen at 20 miles, and the red answering parachute flare fired from the dinghy was seen from a distance of at least 12 miles by the HMS *Opossum*.

Cause of Accident

Although some of the contributing causes remain obscure, the following theory seems to be the best produced to date:

(a) No 2 engine failed and "threw" one or more cylinders due to either failure of the cylinders or holding down bolts. This type of failure is receiving concentrated and detailed study by both the Directorate of Technical Services and the Directorate of Flying Safety.

(b) Failure of No 2 engine also caused the failure of the feathering control to that engine and portion of the engine and/or its cowlings damaged both the airscrew of No 1 engine and the main electric wiring to No 1, or alternatively the extreme vibration from No 2 engine and possibly No 1 engine (which was sufficient to cause the crew grave doubt as to how long the wing or engine mounts would remain intact) may have caused the breaking down of the electrical system and control system to No 1 engine.

(c) Assuming that owing to the break down of the electrical system on the port side, neither the generators of No 1 or No 2 engines were charging the battery or in the case of No 2 engine, the feather motor was operating but the hydraulic lines had failed. The output from the battery which was further taxed by the operation of the W/T transmitter, was sufficiently low to cause extremely sluggish operation of the pitch controls to engines Nos 3 and 4 (a ground test at 1 APU which, although not strictly comparable, showed that with the feather motors on Nos 1 and 2 engines operating, the pitch controls on Nos 3 and 4 had to be held for 15 to 20 seconds before any change of pitch resulted).

Log of Search for A72-348

090235: Message from aircraft to Morotai "may have to ditch".
090240: Distress message from aircraft to Morotai "position 0505N, 12455E – 090235Z., reason for ditching engine trouble".
090243: Message from aircraft "ditching now". Morotai ASR Catalina. NWA and US authorities at Tacloban, Zamboanga and Nichols Field notified by signal.
090315: ASR Catalina 'OLZ' airborne on search.
090454: C-47 Dakota 'CUZ' airborne on search.
090516: Signal from Catalina "Searching from Kawio Island, visibility 5 miles".
090525: Signal from Catalina "Intercept distress message on 500Kcs, have no holding apparatus, inform CUZ".
090528: From Dakota CUZ "Wilco OLZ, intercepting all your messages".
090707: From Dakota CUZ "We are now heading where signal is strongest, OLZ accompanying us".
090735: From Dakota "Picking up SOS fairly strongly, we are trying to home on it". Both aircraft closely searched area where signals were strongest but no trace found. Later signals showed that the area searched was from Kawio Island, distance 60 miles along track, 10 miles either side using visibility 4 miles, then creeping line ahead using visibility 4 miles to Balut Island.
090804: Dakota abandoned search.
090847: Catalina abandoned search.
091100: Catalina ordered to proceed from Darwin to assist search.
091830: Survivors picked up by HMS *Opossum* which had been diverted.

Comments on Search

(a) The visibility of 5 nautical miles used for the search is considered too great as it is generally accepted that the visibility of a dinghy in the sea is only one mile.

(b) As a definite position was given four minutes before ditching and the search aircraft were in the area within three hours of the ditching, it is considered that the correct search procedure would have been to proceed to the estimated dinghy position and carry out a square search using visibility one mile. Receipt of signals from the Gibson Girl should be treated with some reserve until the estimated position of the dinghy is reached owing to the possibility of interfering with the automatic operation of the radio compass.

(c) No explanation, at this stage, can be given for the failure of both aircraft to home on the Gibson Girl. The Gibson Girl transmitted for periods of ten minutes with intervals of ten minutes, and strong signals were received by both aircraft. The captain of the C-47 claimed that his radio compass was serviceable. The captains of the searching aircraft have been requested to conduct a thorough investigation and the results of this investigation will be disseminated in due course.

ROYAL AUSTRALIAN AIR FORCE.

AIRCRAFT ACCIDENT OR FORCED LANDING REPORT.

[To be submitted in accordance with A.F.O's. 10/D 2.]

PART I.—TRAINING.

UNIT No. 23 SQUADRON DATE & TIME 1040 9/4/45 REPORT No.

PLACE CELEBES SEA 05°00'N 124°47'E

AIRCRAFT TYPE LIBERATOR NUMBER A72-348 FLYING HOURS (1) TOTAL 360.30

(2) SINCE LAST ~~XXX~~ 300 HOURLY 57.40

ENGINE TYPE T.B. WASP R1830 - 65A NUMBER(S) & POSITION. RUNNING HOURS (i) TOTAL; (ii) SINCE LAST TOP OVERHAUL

R1830 - 65A	1. 325757 PT OUTER 360.30	
" "	2. 445588 PT INNER 522.00 INSTALED AT 466.20	54.40
" "	3. 326040 STB OUTER ~~INNER~~ 360.30	NEW
" "	4. 452436 STB OUTER 294.25 INSTALLED AT 220.55	NEW

PILOT 1st W/O HANDBURY J.E. 395 DATE GRADUATED FROM FLYING COURSE JUNE 1939

FLYING TIMES—TOTAL 2380 ON TYPE 250 FOR PREVIOUS 12 MONTHS 140

CREW—PILOT 2nd W/O F.L. WELSH (410769)

AIR OBSERVER F/Lt. P.W. MATTHEWS (406080)

W/T OPERATOR W/O C.J. McKENZIE (422241)

OTHERS Radar Op. - W.O. W.N. NESBITT (433943) ENGINEER - W/O R.B. SMITH 19667

ENGINEER - F/Sgt D.W. NORTON 39721 Fitter IIE - Cpl E.E. PYE 34387

PASSENGERS CAPT. J.S. McFADDEN LT. J. POTTEN U.S. SMALL SHIPS.

NATURE OF ACCIDENT:

HEIGHT 9000' No. 2 Engine blew top cylinder & Cowls, would not feather and vibrated badly. Almost simultaneously No. 1 Engine started vibrating seriously & Fluctuated in power. Engine would not feather. Power could not be increased on engines No. 3 & No. 4 aircraft would not maintain height & ditching was carried out.

INJURY TO PERSONNEL:

MINOR INJURIES - W/O HANDBURY W/O NESBITT W/O WELSH W/O SMITH F/S NORTON

SERIOUSLY INJURED - W/O McKENZIE CPL PYE CAPT. McFADDEN

MISSING (Presumed went down with aircraft) F/Lt. MATTHEWS LT. POTTEN.

DAMAGE TO MATERIAL:

COMPLETE WRITE OFF.

REMARKS:

(i) IS COURT OF INQUIRY BEING HELD NO

(ii) HAS DISCIPLINARY ACTION BEEN TAKEN NO

(iii) GENERAL REMARKS FULL REPORT MADE TO D.F.S.

J. Handbury W/Cdr. COMMANDING No 23 SQUADRON.

D.T.S. TECHNICAL NOTE NO.2.

(ISSUE No. 2)

LIBERATOR AIRCRAFT

LOADING DETAILS FOR MODELS B24J, B24L AND B24M WHEN MODIFIED FOR SPECIAL CARGO DUTIES

DIRECTORATE OF TECHNICAL SERVICES

AIR FORCE HEAD-QUARTERS

MELBOURNE, S.C.1.=

OCTOBER, 1945.

(T.S.311/45)

LIBERATOR - FIELD CONVERSION FOR CARRIAGE OF FREIGHT AND/OR PASSENGERS

1. This Technical Note specifies essentially the equipment to be removed and work required to convert training and non operational Liberator aircraft for use as freight or passenger carrying aircraft, having a capacity of 4,000 lbs. when such conversion is ordered by Air Force Head-Quarters.

2. Circumstances may necessitate the employment of aircraft on the establishment of operational units for emergency transport. Authority for their use is to be approved by the A.O.C. of the appropriate Command. When so converted, the items prefixed "non operational aircraft only" are not to be carried out.

3. The nose turret is to remain installed, the guns only to be removed.

4. Ammunition bins and feed chutes in nose turret are to be removed.

5. The stabilizer unit installed forward of Station 0.1 is to be protected adequately with a plywood shelf and vertical metal plate.

6. Non-Operational Aircraft only. At the discretion of the Unit concerned, the mid-upper turret may be removed and the opening framed and skinned over.

7. Operational aircraft only. The mid-upper turret is to remain installed, the guns only may be removed.

8. The port bomb-bay tank only is to remain installed.

9. Install the removable luggage stowages in the rear bomb-bay.

10. Remove the ball turret, and cover over the opening with 3/8 inch aircraft or 1/2 inch commercial plywood, supported by 3 inch x 2 inch wooden transverse bearers bolted at each end to 2 inch x 1/8 inch or 3/16 inch angle brackets resting on the upper edge of the retraction mounting ring.

11. The flooring in the ball turret section of the rear compartment between stations 6 and 7 is to be covered with plywood, the inboard edges to be supported by 3 inch x 2 inch wooden bearers. The plywood floor covering is to extend to the inboard apex of the corrugated floor at Station 7 leaving an opening over the turret well.

12. A vertical plywood panel extending from Stations 6 to 7 and to the height of the command deck, is to be fitted to protect the port side control cable run from being fouled by luggage or personnel.

13. Remove the waist gun installations complete with ammunition bins.

14. <u>Non-Operational Aircraft only</u>. Remove the waist gun armour plating when installed.

15. Cover the corrugated flooring aft of Station 7 with plywood and continue the plywood aprons fixed vertically, below the waist gun window shelf to the floor level.

16. Install a plywood guard over the starboard control cable run between Stations 7.3 and 7.4.

17. Install cargo tying down rings in the following positions:-

(a) Two on the aft face of bulkhead Station 6 on each side of the doorway.

(b) One each on the port and starboard waist window sills immediately forward of the windows.

18. A length of sash cord, if available, or rope approximately 3 ft. long is to be attached to the lower half of each of the two diagonal tubular braces to the side gun post assemblies. A suitable hole is to be cut in the plywood apron to allow for these lengths of cord. This cord is to provide a means of tying the ropes at a rear position, to prevent aft movement of the cargo, during flight.

19. Remove the rear turret ammunition bins.

20. The rear gun installations may be any one of the following three types and will be treated according to the particular installation.

(a) Power operated turret to remain installed, the guns only to be removed.

(b) The hand operated turret is to have the guns and the mounting assembly removed and the turret fixed to prevent rotation.

(c) Where type 6A (Bell) power gun mounts are installed, these are to be removed.

21. <u>Non-Operational Aircraft only</u>. If necessary, a plywood luggage stowage can be built in the front starboard bomb-bay capable of carrying approximately 400 lb. of freight.

22. The aircraft is to have the following loading data painted at the appropriate locations:-

(a)	Bomb aimer's compartment	700 lb.
(b)	Front starboard bomb-bay	400 lb.
(c)	Rear bomb-bay and command deck combined	2400 lb.

(d) Rear compartment stations 6 to 7 — 1000 lb.

(e) Rear compartment stations 7 to 7.4 — 600 lb.

23. The A.N.-A.P.Q.5 removable equipment may be removed.

24. I.F.F. (Mk. III) and Loran installations are to be carried where available.

25. All equipment removed must be carefully stored and immediately reinstalled when the aircraft has completed its temporary transport duties and before despatch from the Unit on the allotment of the aircraft.

26. The number of passengers to be carried is to be in accordance with the directions given in R.A.A.F. Head-Quarters A.774 dated 24th September and A.790 dated 27th September, 1945.

"A.790 As result representations from 11 Group and N.W.A. it has been decided to permit a maximum of 25 repeat 25 passengers to be carried in Liberator aircraft except when evacuating P.O.Ws. or medical cases when 20 repeat 20 personnel only will be carried. It is emphasised that all other limitations imposed by A.774, 24th September remain unaltered."

27. Attached hereto is a weight sheet summary and loading diagram for use in operating and loading Liberators when converted for freight carrying duties.

Group Captain, D.T.S.

I N S T R U C T I O N S

1. Maximum permissible weight = 64,000 lb.

2. C.G. limits = 288.5 to 303.34 inches aft of nose.

3. Maximum loads to be carried in each compartment:-

Compartment A	=	700 lb.
Compartment D (if required)	=	400 lb.
Compartment E x	=	2,400 lb.
Compartment F	=	1,000 lb.
Compartment G	=	600 lb.

4. Maximum dimensions of packages:-

Rear Cargo carrier stowage	=	46"	x 26"	x 16"
Compartments F and G	=	46"	x 36"	x 24"

NOTE: x Compartment E can carry 2,000 lb. in rear bomb bay luggage racks plus 400 lb. on command deck.

LIBERATOR (MODEL B24J)

LOADING DETAILS WHEN CARRYING 4.700 LB. CARGO.

ITEM	WEIGHT (lb.)	ARM (in.)	MOMENT (lb.in.)
CREW (with parachute and dinghy Type K)			
1st Pilot	215	147.0	31,605
2nd Pilot	215	147.0	31,605
Navigator	216	187.8	40,565
Radio operator	216	187.8	40,565
Flight engineer	216	230.0	49,680
RADIO			
Command	75	347.4	26,055
Liaison	230	189.0	43,470
Radio compass	60	367.8	18,390
Marker beacon	3	330.6	992
Interphone	62.6	243.0	15,212
Frequency meter	39	214.2	8,354
I.F.F.	33	280.0	9,240
V.H.F.	90	411.2	37,008
OXYGEN	389	451.4	175,606

ITEM	WEIGHT (lb.)	ARM (in.)	MOMENT (lb.in.)
<u>PYROTECHNICS</u>			
Signal pistol and cartridges (18)	8.5	240.6	2,045
<u>NAVIGATIONAL EQUIPMENT</u>	72	211.1	15,201
<u>MISCELLANEOUS</u>			
Dinghies (2)(Type A.-3) and access.	130	294.6	38,298
Dinghy wireless	33	447.0	14,751
First aid kits (3)	6	247.0	1,482
Vacuum flasks (water)	20	230.0	4,600
Emergency rations	75	230.0	17,250
<u>CARGO</u>			
Compartment A	700	60.0	42,000
Compartment E	2,400	387.0	928,800
Compartment F	1,000	472.0	472,000
Compartment G	600	550.0	330,000
<u>FUEL</u> (at 7.2 lb./gal.)			
Main tanks (1,952 gals.)	14,054	308.0	4,328,632
Auxiliary wing tanks (375 gals.)	2,700	307.0	828,900
<u>OIL</u>			
110 gals. at 9 lb./gal.	990	252.6	250,074
<u>TOTAL REMOVABLE</u>	24,838		
<u>TARE WEIGHT</u> (includes mid-upper turret)	34,548	285.9	9,877,834
<u>TOTALS</u>	59,386	297.7	17,680,214

$\bar{X}$ = 297.7 inches aft of datum.

LIBERATOR (MODEL B24L)

LOADING DETAILS WHEN CARRYING 4,700 LB. CARGO

ITEM	WEIGHT (lb.)	ARM (in.)	MOMENT (lb.in.)
CREW (with parachute and dinghy Type K)			
1st Pilot	215	147.0	31,605
2nd Pilot	215	147.0	31,605
Navigator	216	196.0	42,336
Radio operator	216	196.0	42,336
Flight engineer	216	240.0	51,840
RADIO			
Command	66	350.0	23,100
Liaison	167	200.0	33,400
Radio compass	50	367.8	18,390
Marker beacon	3	330.6	992
Interphone	63	243.0	15,309
Frequency meter	39	220.0	8,580
V.H.F.	90	411.2	37,008
Localizer	24	346.0	8,304
I.F.F.	38	397.3	15,097
SCR.717C	360	235.3	84,708
SCR.729	62	396.8	24,602
AN/APQ5B (may be removed)	90	71.9	6,471
Loran	63	227.9	14,358
Radio altimeter	19	402.0	7,638
OXYGEN	504	560.3	282,408
PYROTECHNICS			
Signal pistol and cartridges (18)	8.5	200.0	1,700
NAVIGATIONAL EQUIPMENT	72	211.1	15,201
MISCELLANEOUS			
Dinghies (2)(Type A-3) and acces.	130	294.6	38,298
Dinghy wireless	34	446.0	15,164
First aid kits (3)	6	247.0	1,482
Vacuum flasks (water)	20	230.0	4,600
Emergency rations	75	230.0	17,250
CARGO			
Compartment A	700	60.0	42,000
Compartment E	2,400	387.0	928,800
Compartment F	1,000	472.0	472,000
Compartment G	600	550.0	330,000

ITEM	WEIGHT (lb.)	ARM (in.)	MOMENT (lb.in.)
FUEL (at 7.2 lb./gal.)			
Main tanks (1,952 gals.)	14,054	308.0	4,328,632
Auxiliary wing tanks (375 gals.)	2,700	307.0	828,900
OIL			
110 gals. at 9 lb./gal.	990	252.6	250,074
TOTAL REMOVABLE	25,506		
TARE WEIGHT (includes mid-upper turret)	35,474	278.0	9,862,377
TOTALS	60,980	293.8	17,916,565

(T.S.311/45) $\bar{x}$ = 293.8 inches aft of datum.

LIBERATOR (MODEL B24M)

LOADING DETAILS WHEN CARRYING 4,700 LB. CARGO

ITEM	WEIGHT (lb.)	ARM (in.)	MOMENT (lb.in.)
CREW (with parachute and dinghy, Type K)			
1st Pilot	215	147.0	31,605
2nd Pilot	215	147.0	31,605
Navigator	216	196.0	42,336
Radio operator	216	196.0	42,336
Flight engineer	216	240.0	51,840
RADIO			
Command	66	350.0	23,100
Liaison	167	200.0	33,400
Radio compass	51	387.0	19,737
Interphone	63	243.0	15,309
Frequency meter	39	220.0	8,580
V.H.F.	90	422.7	38,043
Localizer	24	346.0	8,304
I.F.F.	35	416.0	14,560
SCR.717C	360	235.3	84,708
SCR.729	62	403.2	24,998
AN/APQ5B (may be removed)	90	71.9	6,471
Loran	63	227.9	14,358
Radio altimeter	19	400.0	7,600
Marker beacon	3	359.0	1,077

ITEM	WEIGHT (lb.)	ARM (in.)	MOMENT (lb.in.)
OXYGEN	504	559.5	281,98
PYROTECHNICS			
Signal pistol and cartridges (18)	8.5	200.0	1,70
NAVIGATIONAL EQUIPMENT	72	211.1	15,20
MISCELLANEOUS			
Dinghies (2)(Type A-3) and access.	130	294.6	38,29
Dinghy wireless	34	476.0	16,18
First aid kits (3)	6	247.0	1,48
Vacuum flasks (water)	20	230.0	4,60
Emergency rations	75	230.0	17,25
CARGO			
Compartment A	700	60.0	42,00
Compartment E	2,400	387.0	928,80
Compartment F	1,000	472.0	472,00
Compartment G	600	550.0	330,00
FUEL (at 7.2 lb./gal.)			
Main tanks (1,952 gals.)	14,054	308.0	4,328,63
Auxiliary wing tanks (375 gals.)	2,700	307.0	828,90
OIL			
110 gals. at 9 lb./gal.	990	252.6	250,07
TOTAL REMOVABLE	25,504		
TARE WEIGHT (includes mid-upper turret)	35,604	281.6	10,024,86
TOTALS	61,108	295.9	18,081,94

$\bar{x}$ = 295.9 inches aft of datum.

(T.S.311/45)

LIBERATOR (MODEL B24M)

LOADING DETAILS WHEN CARRYING 25 PASSENGERS.

ITEM	WEIGHT (lb.)	ARM (in.)	MOMENT (lb.in.)
CREW (with parachute and dinghy, Type K)			
1st Pilot	215	147.0	31,605
2nd Pilot	215	147.0	31,605
Navigator	216	196.0	42,336
Radio operator	216	196.0	42,336
Flight engineer	216	240.0	51,840
RADIO			
Command	66	350.0	23,100
Liaison	167	200.9	33,400
Radio compass	51	387.0	19,737
Interphone	63	243.0	15,309
Frequency meter	39	220.0	8,580
V.H.F.	90	422.7	38,043
Localizer	24	346.0	8,304
I.F.F.	35	416.0	14,560
SCR.717C	360	235.3	84,708
SCR.729	62	403.2	24,998
AN/APQ5B (may be removed)	90	71.9	6,471
Loran	63	227.9	14,358
Radio altimeter	19	400.0	7,600
Marker beacon	3	359.0	1,077
OXYGEN	504	559.5	281,988
PYROTECHNICS			
Signal pistol and cartridges (18)	8.5	200.0	1,700
NAVIGATIONAL EQUIPMENT	72	211.1	15,201
MISCELLANEOUS			
Dinghies (2)(Type A-3) and access.	130	294.6	38,298
Dinghy wireless	34	476.0	16,184
Vacuum flasks (water)	20	230.0	17,250
Luggage (Compartment D)	750	280.0	210,000
PASSENGERS (at 180 lb. each)			
Compartment E:-			
Bomb bay (6)	1,080	387.0	417,960
Command deck (3)	540	387.0	208,980
Compartment F (6)	1,080	472.0	509,760
Compartment G (10)	1,800	550.0	990,000

ITEM	WEIGHT (lb.)	ARM (in.)	MOMENT (lb.in.)
FUEL (at 7.2 lb./gal.)			
Main tanks (1,952 gals.)	14,054	308.0	4,328,632
Auxiliary wing tank (375 gals.)	2,700	307.0	828,900
OIL			
110 gals. at 9 lb./gal.	990	252.6	250,074
TOTAL REMOVABLE	25,973		
TARE WEIGHT	35,604	281.6	10,024,86[illegible]
TOTALS	61,577	302.5	18,639,761

x = 302.5 inches aft of datum

(T.S.311/45)

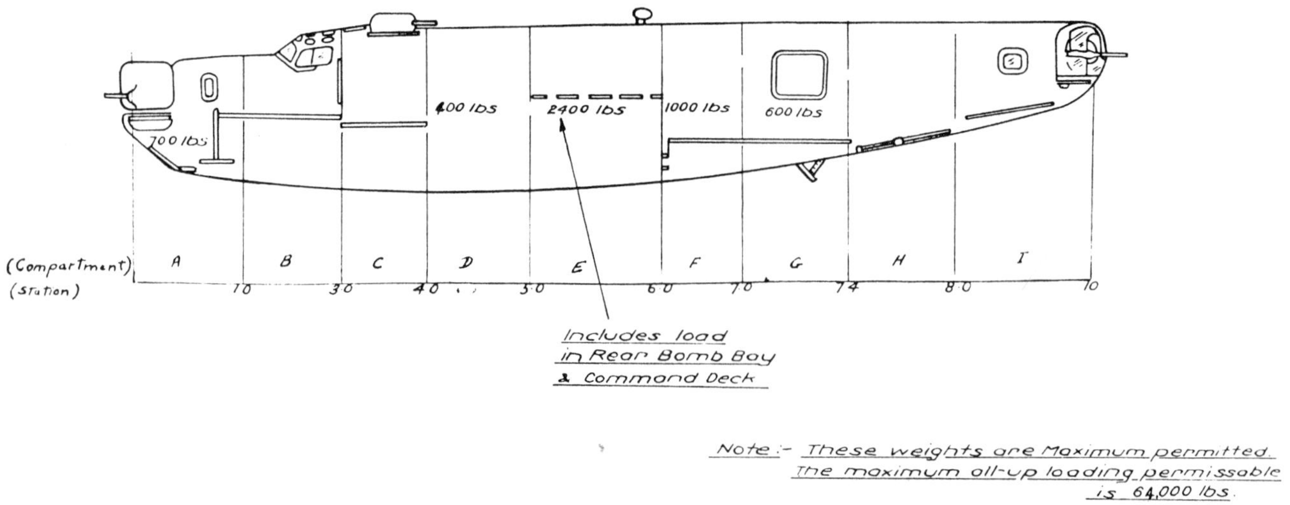

LOADING B24 AIRCRAFT:- SPECIAL TRANSPORT

A 8905

RAAF CONSOLIDATED LIBERATORS

Abbreviations: w/o – written off; wfu – withdrawn from use; soc – struck off charge; cvtd – converted; compnts – components; deliv – delivery; t/o – takeoff; n/a – not applicable/available; ops – operations; ftr – failed to return; dbr – damaged beyond repair; USAAF – United Staes Army Air Force; DCA – Department of Civil Aviation.

RAAF No	Model	USAAF No	Deliv	Disposal/Remarks
A72-1	B-24D	41-11904	02/44	damaged in fire Tocumwal NSW 10/44, cvtd to compnts 03/48
A72-2	B-24D	41-23720	02/44	stored 03/46, soc 12/52
A72-3	B-24D	41-24018	02/44	soc 03/48
A72-4	B-24D	42-40489	02/44	damaged in fire Tocumwal NSW 10/44, cvtd to compnts (7 OTU)
A72-5	B-24D	42-40512	02/44	stored 03/46, soc 06/50
A72-6	B-24D	42-40522	02/44	cvtd to components 08/47
A72-7	B-24D	41-124070	02/44	cvtd to components 10/45
A72-8	B-24D	41-24290	02/44	crashed on t/o Tocumwal NSW 01/45 (7OTU)
A72-9	B-24D	41-11868	02/44	cvtd to components 09/45
A72-10	B-24D	41-24127	05/44	soc 12/52
A72-11	B-24D	42-40514	07/44	emergency landing Bankstown NSW 04/45, cvtd to compnts 11/45
A72-12	B-24D	42-41132	07/44	returned to USAAF 04/45
A72-13	B-24J	42-100194	07/44	returned to USAAF 04/45
A72-14 to A72-30 not allocated				
A72-31	B-24J	44-40251	05/44	cvtd to components 03/46
A72-32	B-24J	44-40252	06/44	soc 12/52
A72-33	B-24J	44-40410	06/44	soc 12/52
A72-34	B-24J	44-40406	05/44	soc 12/52
A72-35	B-24J	44-40408	06/44	soc 12/52
A72-36	B-24J	44-40409	06/44	crashed on landing Tocumwal NSW 04/45 (7 OTU)
A72-37	B-24J	44-40407	05/44	soc 12/52
A72-38	B-24J	44-40411	06/44	soc 12/52
A72-39	B-24J	44-40653	06/44	landing accident 09/44, cvtd to compnts (24 Sqdn)
A72-40	B-24J	44-40649	06/44	soc 07/52
A72-41	B-24J	44-40652	06/44	soc 12/52
A72-42	B-24J	44-40651	06/44	soc 12/52
A72-43	B-24J	44-40654	06/44	soc 12/52
A72-44	B-24J	44-40650	06/44	soc 12/52
A72-45	B-24J	44-40870	07/44	soc 07/52
A72-46	B-24J	44-40871	07/44	soc 12/52
A72-47	B-24J	44-40872	07/44	soc 07/52
A72-48	B-24J	44-40873	07/44	soc 12/52
A72-49	B-24J	44-41194	08/44	authorised for w/o 12/46, sold 06/50
A72-50	B-24J	44-41197	08/44	soc 12/52
A72-51	B-24J	44-41203	08/44	soc 12/52
A72-52	B-24J	44-41207	09/44	soc 12/52
A72-53	B-24J	44-41204	08/44	authorised for w/o 12/46, sold 06/50
A72-54	B-24J	44-41196	08/44	soc 12/52
A72-55	B-24J	44-41206	08/44	soc 12/52
A72-56	B-24J	44-41193	08/44	authorised for w/o 12/46, sold 06/50
A72-57	B-24J	44-41205	08/44	soc 12/52
A72-58	B-24J	44-41195	08/44	soc 12/52
A72-59	B-24J	44-41389	09/44	damaged by enemy action 03/45, cvtd to compnts (24 Sqdn)
A72-60	B-24J	44-41388	09/44	soc 12/52
A72-61	B-24J	44-41375	09/44	lost to enemy action Balikpapan area 07/45 (21 Sqdn)
A72-62	B-24J	44-41249	10/44	soc 12/52
A72-63	B-24J	44-41374	10/44	soc 12/52
A72-64	B-24J	44-41384	09/44	lost to enemy action Balikpapan area 07/45 (24 Sqdn)
A72-65	B-24J	44-41385	09/44	soc 12/52
A72-66	B-24J	44-41373	09/44	ftr from op 08/45 (21 Sqdn)

RAAF No	Model	USAAF No	Deliv	Disposal/Remarks
A72-67	B-24J	44-41386	09/44	authorised for w/o 12/46, sold 06/50
A72-68	B-24J	44-41376	09/44	soc 12/52
A72-69	B-24L	44-41391	09/44	soc 12/52
A72-70	B-24L	44-41392	09/44	missing on ops 01/45 (24 Sqdn)
A72-71	B-24L	44-41393	09/44	soc 02/52
A72-72	B-24L	44-41394	09/44	authorised for w/o 12/46, sold 06/50
A72-73	B-24L	44-41401	09/44	authorised for w/o 12/46, sold 06/50
A72-74	B-24L	44-41395	09/44	soc 12/52
A72-75	B-24L	44-41403	10/44	soc 12/52
A72-76	B-24L	44-41402	09/44	soc 12/52
A72-77	B-24L	44-41404	09/44	shot down near Koepang 04/45 (24 Sqdn)
A72-78	B-24L	44-41405	09/44	soc 12/52
A72-79	B-24L	44-41444	10/44	authorised for w/o 12/46, sold 06/50
A72-80	B-24L	44-41450	10/44	crashed into Flores Sea 03/45 (24 Sqdn)
A72-81	B-24L	44-14154	10/44	shot down Casualty Point 04/45 (24 Sqdn)
A72-82	B-24L	44-41452	10/44	soc 12/52
A72-83	B-24L	44-41455	10/44	soc 08/49
A72-84	B-24L	44-41456	10/44	soc 12/52
A72-85	B-24L	44-41457	10/44	soc 08/49
A72-86	B-24L	44-41458	10/44	soc 12/52
A72-87	B-24L	44-41459	10/44	soc 08/49
A72-88	B-24L	44-41460	10/44	crashed near Fenton NT 02/45 (24 Sqdn)
A72-89	B-24I	44-41575	10/44	soc 12/52
A72-90	B-24L	44-41579	10/44	sold 06/50
A72-91	B-24L	44-41580	10/44	soc 12/52
A72-92	B-24L	44-41581	10/44	ftr search mission 07/45 (21 Sqdn)
A72-93	B-24L	44-41582	10/44	soc 12/52
A72-94	B-24L	44-41583	10/44	sold 06/50
A72-95	B-24L	44-41604	10/44	soc 12/52
A72-96	B-24L	44-41653	10/44	soc 12/52
A72-97	B-24L	44-41608	10/44	sold 06/50
A72-98	B-24L	44-41639	10/44	sold 06/50
A72-99	B-24L	44-41640	10/44	soc 12/52
A72-100	B-24L	44-41609	10/44	soc 12/52
A72-101	B-24L	44-41654	10/44	soc 12/52
A72-102	B-24L	44-41611	10/44	soc 12/52
A72-103	B-24L	44-41577	10/44	soc 12/52
A72-104	B-24L	44-41612	10/44	soc 12/52
A72-105	B-24L	44-41617	10/44	soc 12/52
A72-106	B-24L	44-41605	10/44	sold 06/50
A72-107	B-24L	44-41607	10/44	soc 12/52
A72-108	B-24L	44-41618	10/44	soc 12/52
A72-109	B-24L	44-41630	11/44	sold 06/50
A72-110	B-24L	44-41656	11/44	soc 12/52
A72-111	B-24L	44-41663	11/44	soc 12/52
A72-112	B-24L	44-41664	11/44	crashed Tocumwal NSW 02/45 (7 OTU)
A72-113	B-24L	44-41629	11/44	soc 12/52
A72-114	B-24L	44-41633	11/44	soc 12/52
A72-115	B-24L	44-41632	11/44	sold 06/50
A72-116	B-24L	44-41631	11/44	soc 12/52
A72-117	B-24L	44-41634	11/44	soc 12/52
A72-118	B-24L	44-41635	11/44	soc 12/52
A72-119	B-24L	44-41636	11/44	soc 12/52
A72-120	B-24L	44-41642	11/44	soc 12/52
A72-121	B-24L	44-41666	11/44	soc 12/52
A72-122	B-24L	44-41637	11/44	soc 03/48
A72-123	B-24L	44-41638	11/44	crash landed Tocumwal NSW 01/45, cvtd to compnts (7 OTU)
A72-124	B-24L	44-41657	12/44	crashed on t/o Cunderdin WA 02/45 (25 Sqdn)

RAAF No	Model	USAAF No	Deliv	Disposal/Remarks
A72-125	B-24L	44-41505	11/44	soc 12/52
A72-126	B-24L	44-41499	11/44	soc 12/52
A72-127	B-24L	44-41502	11/44	soc 12/52
A72-128	B-24L	44-41503	11/44	converted to components 10/45
A72-129	B-24L	44-41504	11/44	soc 12/52
A72-130	B-24L	44-41506	11/44	soc 12/52
A72-131	B-24L	44-41508	11/44	soc 12/52
A72-132	B-24L	44-41507	11/44	soc 12/52
A72-133	B-24L	44-41509	11/44	ditched returning from op 04/45 (25 Sqdn)
A72-134	B-24L	44-41510	11/44	taxying accident Cunderdin WA 03/45, cvtd to compnts (25 Sqdn)
A72-135	B-24L	44-41511	12/44	soc 12/52
A72-136	B-24L	44-41512	11/44	soc 12/52
A72-137	B-24L	44-41513	11/44	dbr 11/45, cvtd to compnts (25 Sqdn)
A72-138	B-24L	44-41514	11/44	soc 12/52
A72-139	B-24L	44-41658	11/44	sold 06/50
A72-140	B-24L	44-41665	11/44	soc 12/52
A72-141	B-24L	44-41677	11/44	soc 12/52
A72-142	B-24L	44-41682	11/44	soc 12/52
A72-143	B-24M	44-41828	11/44	soc 12/52
A72-144	B-24M	44-41831	11/44	soc 12/52
A72-145	B-24M	44-41834	12/44	soc 12/52
A72-146	B-24M	44-41836	12/44	soc 12/52
A72-147	B-24M	44-41885	12/44	sold 06/50
A72-148	B-24M	44-41886	01/45	soc 12/52
A72-149	B-24L	44-41515	11/44	soc 06/48
A72-150	B-24L	44-41516	12/44	soc 12/52
A72-151	B-24L	44-41517	11/44	soc 12/52
A72-152	B-24L	44-41518	11/44	soc 12/52
A72-153	B-24L	44-41530	11/44	soc 12/52
A72-154	B-24L	44-41531	12/44	sold 06/50
A72-155	B-24L	44-41532	11/44	soc 12/52
A72-156	B-24L	44-41533	12/44	soc 12/52
A72-157	B-24L	44-41529	12/44	soc 12/52
A72-158	B-24M	44-41971	01/45	soc 12/52
A72-159	B-24M	44-41972	01/45	missing on ops 05/45 (200 Flt)
A72-160	B-24M	44-41949	12/44	crashed on t/o 05/45 (12 Sqdn)
A72-161	B-24M	44-41950	01/45	converted to components 05/47
A72-162	B-24M	44-41951	01/45	soc 12/52
A72-163	B-24M	44-41952	01/45	soc 12/52
A72-164	B-24M	44-41953	01/45	soc 12/52
A72-165	B-24M	44-41954	12/44	soc 12/52
A72-166	B-24M	44-41955	01/45	to DCA 12/49 for fire training
A72-167	B-24M	44-41957	01/45	sold 06/50
A72-168	B-24M	44-41958	01/45	soc 12/52
A72-169	B-24M	44-41959	01/45	soc 12/52
A72-170	B-24M	44-41960	01/45	soc 12/52
A72-171	B-24M	44-41961	01/45	soc 12/52
A72-172	B-24M	44-41962	01/45	sold 06/50
A72-173	B-24M	44-41963	01/45	soc 12/52
A72-174	B-24M	44-41964	01/45	soc 12/52
A72-175	B-24M	44-41965	01/45	soc 12/52
A72-176	B-24M	44-41956	01/45	soc 03/48, fuselage extant with G Toye, Moe Vic
A72-177	B-24M	44-41966	02/45	lost on ops 05/45 (200 Flt)
A72-178	B-24M	44-41967	01/45	soc 12/52
A72-179	B-24M	44-41969	01/45	soc 12/52
A72-180	B-24M	44-41970	01/45	sold 06/50
A72-181	B-24M	44-41973	01/45	soc 12/52
A72-182	B-24M	44-41974	01/45	soc 12/52
A72-183	B-24M	44-41975	02/45	sold 06/50
A72-184	B-24M	44-41976	01/45	converted to components 09/45
A72-185	B-24M	44-41977	01/45	sold 06/50
A72-186	B-24M	44-41978	01/45	to DCA 12/51 for fire training Essendon Vic
A72-187	B-24M	44-41979	01/45	soc 12/52
A72-188	B-24M	44-41980	01/45	soc 12/52
A72-189	B-24M	44-41982	01/45	sold 06/50
A72-190	B-24M	44-41983	01/45	soc 12/52
A72-191	B-24M	44-41984	01/45	missing on ops 03/45 (200 Flt)
A72-192	B-24M	44-41987	01/45	sold 06/50
A72-193	B-24M	44-41988	01/45	sold 06/50
A72-194	B-24M	44-41989	01/45	soc 12/52
A72-195	B-24M	44-41990	01/45	soc 12/52

RAAF No	Model	USAAF No	Deliv	Disposal/Remarks
A72-196	B-24M	44-41991	02/45	shot down Balikpapan area 06/45 (24 Sqdn)
A72-197	B-24M	44-41992	02/45	to DCA 12/52 for fire training Essendon Vic
A72-198	B-24M	44-41993	01/45	soc 12/52
A72-199 to A72-299 not allocated				
A72-300	B-24J	44-28063	03/45	soc 05/52
A72-301	B-24J	44-28064	04/45	soc 12/52
A72-302	B-24J	44-28065	04/45	soc 05/52
A72-303	B-24J	44-28066	04/45	soc 05/52
A72-304	B-24J	44-28067	04/45	soc 05/52
A72-305	B-24J	44-28069	04/45	soc 05/52
A72-306	B-24J	44-28087	04/45	crashed on t/o Amberley Qld 08/45 (99 Sqdn)
A72-307	B-24J	44-28097	04/45	soc 05/52
A72-308	B-24J	44-28099	04/45	soc 12/52
A72-309	B-24J	44-28101	04/45	soc 05/52
A72-310	B-24J	44-28104	04/45	soc 05/52
A72-311	B-24J	44-28139	04/45	soc 05/52
A72-312	B-24J	44-28086	04/45	soc 05/52
A72-313	B-24J	44-28125	04/45	crashed 09/45 (99 Sqdn)
A72-314	B-24J	44-28127	04/45	damaged by enemy action 07/45, cvtd to compnts (12 Sqdn)
A72-315	B-24J	44-28256	04/45	to DCA 12/51 for fire training Essendon Vic
A72-316	B-24J	44-28079	04/45	soc 05/52
A72-317	B-24J	44-28122	04/45	engine fire after t/o 09/45, cvtd to compnts (12 Sqdn)
A72-318	B-24J	44-28070	04/45	soc 12/52
A72-319	B-24J	44-28082	04/45	soc 05/52
A72-320	B-24J	44-28132	04/45	soc 05/52
A72-321	B-24J	44-28089	04/45	soc 05/52
A72-322	B-24J	44-28076	04/45	sold 06/50
A72-323	B-24J	44-28103	04/45	soc 05/52
A72-324	B-24J	44-28138	05/45	soc 05/52
A72-325	B-24J	44-28107	05/45	soc 05/52
A72-326	B-24J	44-28118	05/45	soc 05/52
A72-327	B-24J	44-28123	05/45	to DCA 03/51 for fire training
A72-328	B-24J	44-28092	05/45	soc 05/52
A72-329	B-24J	44-28100	05/45	soc 12/52
A72-330	B-24J	44-28114	05/45	soc 05/52
A72-331	B-24J	44-28068	05/45	soc 05/52
A72-332	B-24J	44-28076	05/45	soc 05/52
A72-333	B-24J	44-28098	05/45	soc 05/52
A72-334	B-24J	44-28091	05/45	soc 12/52
A72-335	B-24J	44-28081	05/45	soc 12/52
A72-336	B-24J	44-28108	05/45	soc 12/52
A72-337	B-24J	44-28078	05/45	soc 05/52
A72-338	B-24J	44-28095	05/45	soc 05/52
A72-339	B-24J	44-28124	05/45	soc 07/52
A72-340	B-24J	44-28096	05/45	soc 07/52
A72-341	B-24J	44-28113	05/45	soc 07/52
A72-342	B-24J	44-28137	05/45	soc 07/52
A72-343	B-24J	44-28080	05/45	soc 05/52
A72-344	B-24J	44-28106	05/45	soc 07/52
A72-345	B-24J	44-28084	05/45	soc 07/52
A72-346	B-24J	44-28126	05/45	soc 12/52
A72-347	B-24J	44-28109	05/45	soc 12/52
A72-348	B-24J	44-28072	05/45	ditched into Celebes Sea 04/46 Manila/Darwin travel flight (23 Sqdn)
A72-349	B-24J	44-28077	05/45	soc 12/52
A72-350	B-24J	44-28090	05/45	soc 07/52
A72-351	B-24J	44-28074	05/45	soc 05/52
A72-352	B-24J	44-28135	05/45	soc 05/52
A72-353	B-24J	44-28133	06/45	soc 05/52
A72-354	B-24J	44-28083	06/45	soc 05/52
A72-355	B-24J	44-28085	06/45	soc 05/52
A72-356	B-24J	44-28130	06/45	soc 05/52
A72-357	B-24J	44-28129	06/45	soc 05/52
A72-358	B-24J	44-28093	06/45	soc 05/52
A72-359	B-24J	44-28111	06/45	crashed on t/o Darwin NT 11/45 (23 Sqdn)
A72-360	B-24J	44-28120	06/45	soc 10/51
A72-361	B-24J	44-28134	06/45	soc 05/52
A72-362	B-24J	44-28073	06/45	soc 05/52

RAAF No	Model	USAAF No	Deliv	Disposal/Remarks
A72-363	B-24J	44-28156	06/45	soc 05/52
A72-364	B-24J	44-28168	06/45	soc 05/52
A72-365	B-24J	44-28156	06/45	soc 05/52
A72-366	B-24J	42-78787	06/45	sold 06/50
A72-367	B-24J	44-28088	06/45	soc 05/52
A72-368	B-24J	44-28094	06/45	soc 05/52
A72-369	B-24J	44-28136	06/45	sold Wilmore Aviation, Matraville NSW 12/52
A72-370	B-24J	44-28144	06/45	soc 05/52
A72-371	B-24J	44-28164	06/45	soc 05/52
A72-372	B-24J	44-28140	06/45	soc 05/52
A72-373	B-24J	44-28148	06/45	soc 05/52
A72-374	B-24J	44-28071	06/45	soc 05/52
A72-375	B-24J	44-28102	06/45	soc 05/52
A72-376	B-24J	44-28121	06/45	soc 05/52
A72-377	B-24J	44-28174	06/45	soc 12/52
A72-378	B-24J	44-28171	06/45	soc 12/52
A72-379	B-24J	44-28143	06/45	soc 05/52
A72-380	B-24J	44-28145	06/45	soc 12/52
A72-381	B-24J	44-28151	06/45	instructional airframe 07/47, soc 05/53
A72-382	B-24J	44-28183	06/45	soc 12/52
A72-383	B-24J	44-28142	06/45	to DCA Essendon Vic 12/51 for fire training
A72-384	B-24J	44-28162	06/45	soc 05/52
A72-385	B-24J	44-28179	06/45	soc 05/52
A72-386	B-24J	44-28155	07/45	soc 05/52
A72-387	B-24J	44-28062	07/45	soc 12/52
A72-388	B-24J	44-28169	07/45	soc 05/52
A72-389	B-24J	44-28167	07/45	soc 05/52
A72-390	B-24J	44-28166	07/45	soc 12/52
A72-391	B-24J	44-28146	07/45	soc 12/52
A72-392	B-24J	44-28178	07/45	soc 12/52
A72-393	B-24J	44-28075	07/45	soc 07/52
A72-394	B-24J	44-28159	07/45	intended for conversion to VIP config but cancelled, soc 07/52
A72-395	B-24J	44-28180	07/45	soc 07/52
A72-396	B-24J	44-28172	07/45	soc 07/52
A72-397	B-24J	44-28175	07/45	soc 05/52
A72-398	B-24J	44-28158	07/45	soc 07/52
A72-399	B-24J	44-28150	07/45	soc 07/52
A72-400	B-24J	44-28147	08/45	soc 12/52
A72-401	B-24J	44-28153	08/45	soc 05/52
A72-402	B-24J	44-28160	08/45	soc 12/52
A72-403	B-24J	44-28161	08/45	soc 12/52
A72-404	B-24J	42-78726	08/45	soc 12/52
A72-405	B-24J	42-78727	08/45	soc 12/52

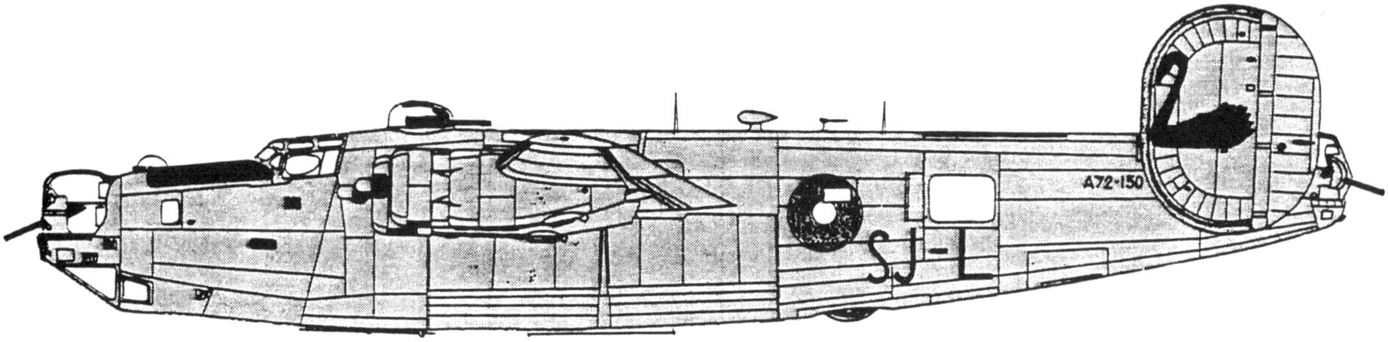

B-42L A72-150/SJ-L of No 25 Sqn
Natural metal overall. Black code, serial and swan on fin. Olive drab anti-glare panel in front of cockpit. No roundels under the wings.

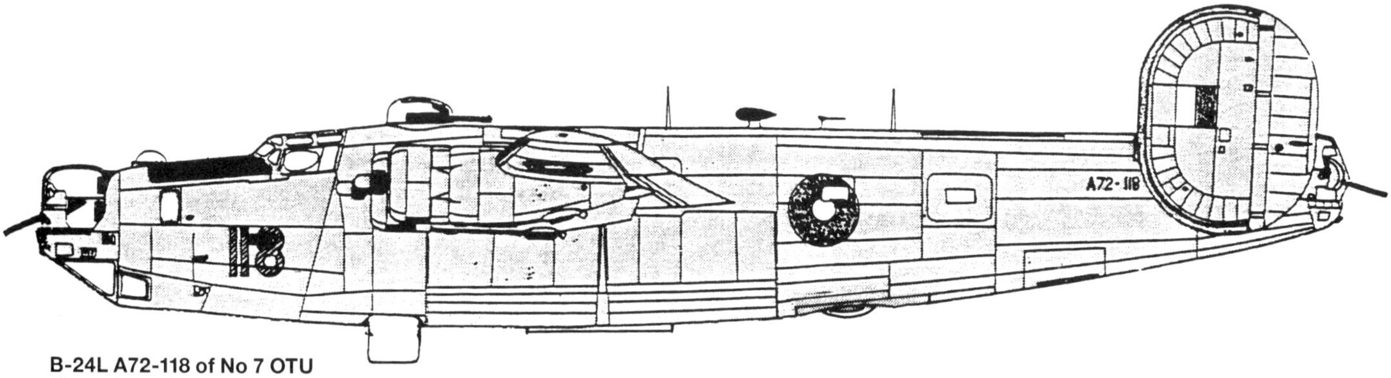

B-24L A72-118 of No 7 OTU
Natural metal overall. Black serial. Olive drab anti-glare panel in front of cockpit. Yellow numbers with thin black outline on nose. White radar dome (shown extended) below forward fuselage.

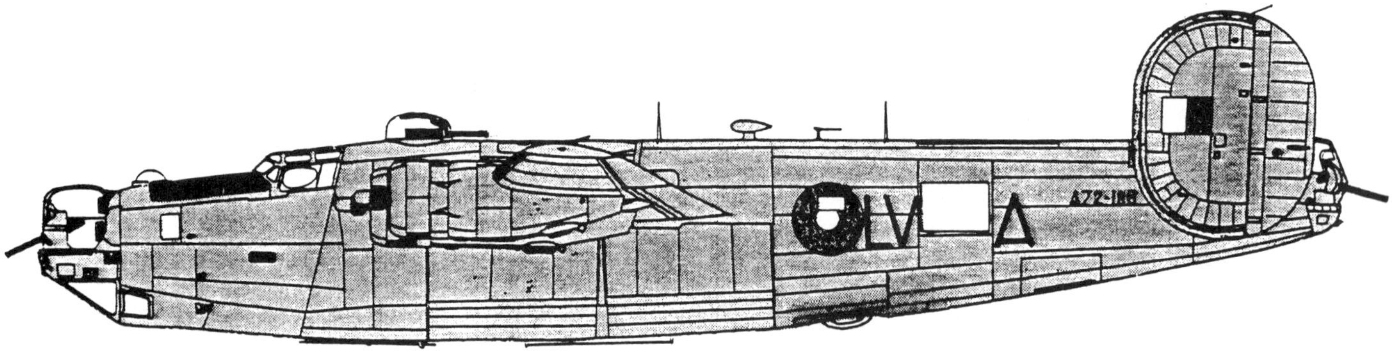

B-24M A72-186/LV-A of 201 Flight
Natural metal overall. Black code and serial. Olive drab anti-glare panel in front of cockpit. No roundels under the wings.

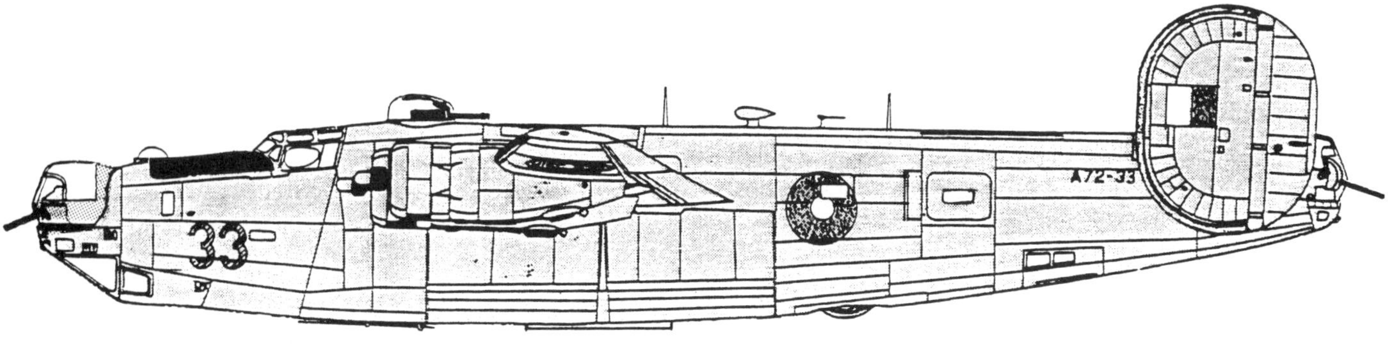

B-24J A72-33 of No 7 OTU
Natural metal overall. Black serial. Olive drab anti-glare panel in front of cockpit. Yellow numbers with thin black outline on nose.

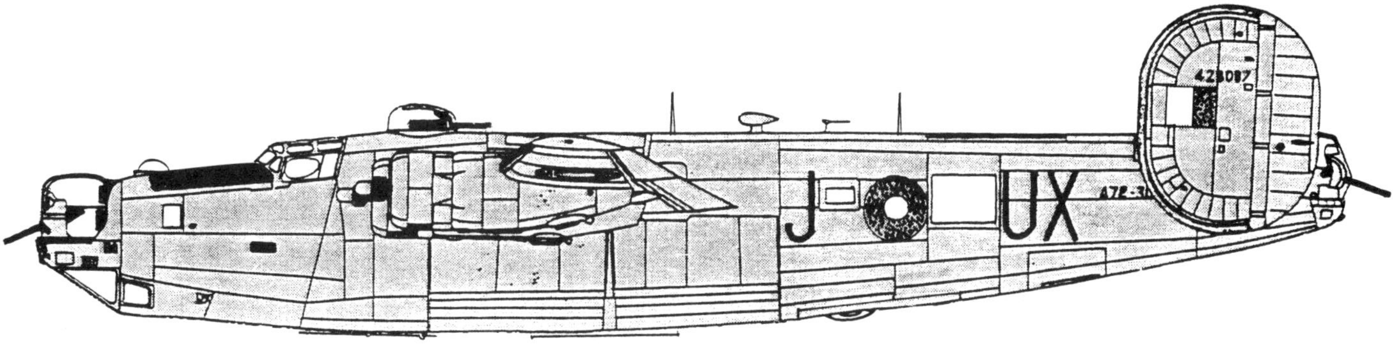

B-24J A72-306/UX-J of No 99 Sqn
Natural metal overall. Black code and serial. Olive drab anti-glare panel in front of cockpit.

B-24J A72-56/MJ-W of No 21 Sqn
Natural metal overall. Black code, serial and chevron on fin. Olive drab anti-glare panel in front of cockpit. Five bombing mission marks below cockpit in black.

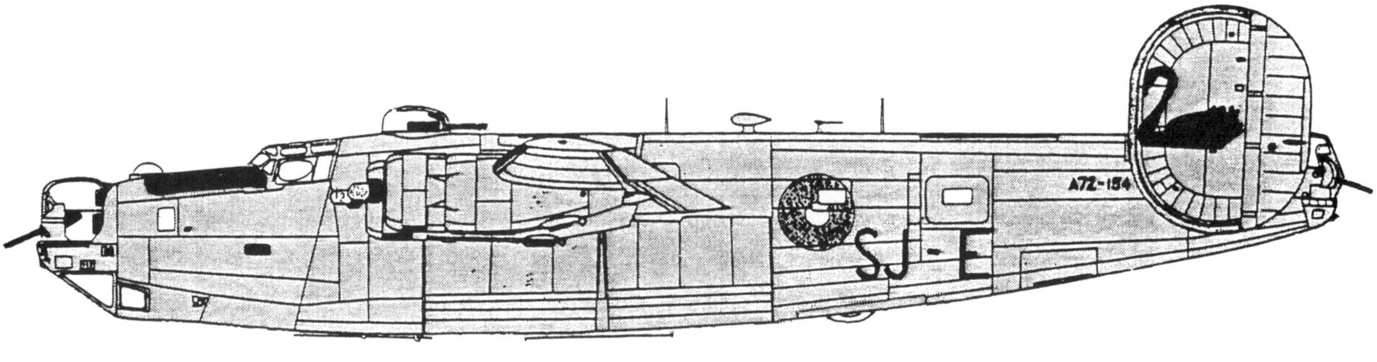

B-24L A72-154/SJ-E of No 25 Sqn
Natural metal overall. Black code, serial and swan on fin. Olive drab anti-glare panel in front of cockpit. No roundels under the wings.

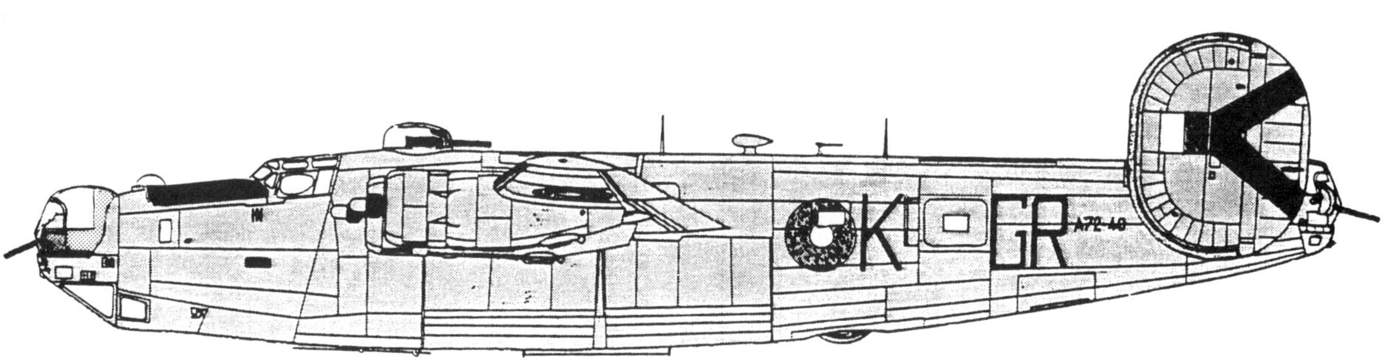

B-24J A72-40/GR-K of No 24 Sqn
Natural metal overall. Black code and serial. Olive drab anti-glare panel in front of cockpit. Three bombing mission marks below cockpit in black.

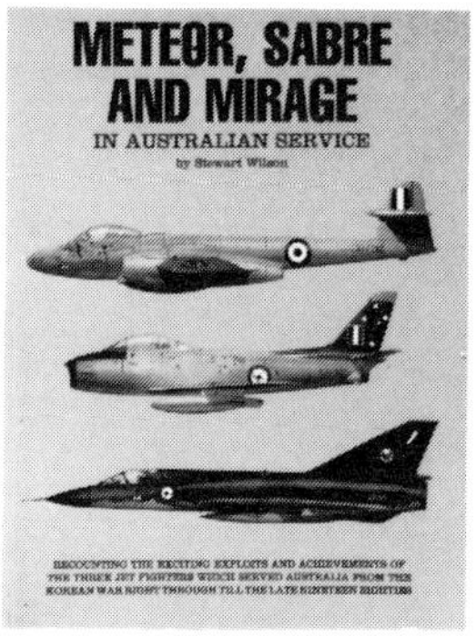

Have you missed out on earlier editions of our unique AUSTRALIAN AIRPOWER COLLECTION series?

If so, order your copy now!

POST TO: AEROSPACE PUBLICATIONS PTY LTD
PO BOX 3105, WESTON CREEK 2611
TEL: (06) 288 1677, FAX: (06) 288 2021

Soft cover $19.95   Hard cover $29.95
(Please add $4.95 per item for postage)
($A30 or $A40 respectively for overseas seamail orders).

Soft	Hard	
☐	☐	Spitfire, Mustang & Kittyhawk
☐	☐	Meteor, Sabre & Mirage
☐	N/A	Lincoln, Canberra & F-111
☐	N/A	Beaufort, Beaufighter & Mosquito
☐	N/A	Dakota, Hercules & Caribou
☐	☐	Catalina, Neptune & Orion
☐	☐	Wirraway, Boomerang & CA-15

☐ I enclose my cheque/MO for $ .... payable to Aerospace Publications P/L or please charge to:

☐ Bankcard ☐ MasterCard ☐ AmEx ☐ Visa

Card No: ....

Expiry Date: .... Ph: ....

Signature ....

NAME ....

ADDRESS ....

.... P/C ....

Please allow 3 weeks for delivery

Future books include the Anson, Hudson & Sunderland and the Vampire, Macchi & PC-9.

A SERIES NOT TO BE MISSED!